MELVILLE IN LOVE

ALSO BY MICHAEL SHELDEN

Friends of Promise:
Cyril Connolly and the World of Horizon

Orwell: The Authorized Biography

Graham Greene: The Enemy Within

Mark Twain: Man in White,
The Grand Adventure of His Final Years

Young Titan: The Making of Winston Churchill

MELVILLE IN LOVE

THE SECRET LIFE OF HERMAN MELVILLE AND THE MUSE OF *MOBY-DICK*

MICHAEL SHELDEN

ecce

An Imprint of HarperCollins*Publishers*

HarperCollins books may be purchased for educational,
business, or sales promotional use. For information please e-mail
the Special Markets Department at SPsales@harpercollins.com.

FIRST EDITION

Designed by Ashley Tucker
Sea art by Marzolino/Shutterstock, Inc.

Library of Congress Cataloging-in-Publication Data
has been applied for.

ISBN 978-0-06-241898-2

16 17 18 19 20   OV/RRD   10 9 8 7 6 5 4 3 2 1

TO SUE

AND TO THE MEMORY OF

L. W. & J. N. MITCHELL

Most considerate of all the delicate roses that diffuse their blessed perfume among men, is Mrs. Morewood.

—HERMAN MELVILLE TO SARAH MOREWOOD, 1851

CONTENTS

THE LAUREL WREATH

It is a winter scene like one you'd find in a faded calendar of old New England. A snowstorm has swept through the countryside overnight, sending the temperature plummeting and turning the landscape white. And there, on a gentle rise in a valley between two mountains in the Berkshires, is an old colonial mansion with plumes of smoke rising from its tall chimneys, and guests arriving in sleighs for Christmas dinner.

The year is 1851, and the house is the pride of Pittsfield, the nearest town on this western edge of Massachusetts. Lately christened "Broadhall" by its new owners, the elegant mansion has two grand parlors with fancy chandeliers, separated by a wide hall and a solid old staircase. It was built by expert craftsmen using huge oak timbers from the area, and the upper windows command a sweeping view of the snow-covered mountains and fields that spread out in all directions. To the north, about a mile away, the church towers of the little town dot the horizon.

At the door servants usher inside the dozen or so guests, all from the neighborhood, including a doctor, a rich farmer, the town historian, and their various wives and daughters. The parlors glow with candles and crackling fires, the table is set, and decorations are everywhere for "an old-fashion'd English Christmas, with Holly & Mistletoe, & bobbing apples."[1]

The hostess for this gathering is Sarah Anne Morewood, the twenty-eight-year-old wife of an English-born merchant and trader. Her husband is the prosperous but bland John Rowland Morewood, who also keeps a house in Manhattan, 150 miles away, where he devotes much of his time to his business and his local Episcopal church. His pretty wife likes living in the country. When the weather allows, she delights in exploring the Berkshire scenery on long "rambles" of ten miles on foot, or on rides of twenty miles on horseback. A free spirit who enjoys defying convention, she has a direct and open manner that can be unnerving, and she is full of strong passions. Proudly, she tells friends, "My feelings . . . are always intense." The few surviving pictures of her capture that intensity in her eyes, which are dark and penetrating.[2]

In solitary hours she is sometimes known to take paper and pencil to the woods and write lyrical verses about nature and love and death and other subjects typical of so many poems of the time. The editor of the *Pittsfield Sun* is an admirer of her writing, and a local church choir has set one of her verses to music. Thanks to her wealth, her literary interests, and her strong personality, she is already a prominent figure in the community, though her independent ways have also generated considerable gossip. It is whispered that she has shown too much interest in other men when Mr. Morewood is away.

Her husband is present on this Christmas afternoon, but Sarah's interest is, in fact, focused on a male guest at her party. He is the owner of the adjoining farm, an author who has just published a sprawling novel about the doomed pursuit of a great white whale across the distant reaches of the Pacific.

OVER THE PAST YEAR, in an upper room of his farmhouse overlooking this rolling countryside, Herman Melville has completed *Moby-Dick; or, the Whale,* and the book is now in print. Landlocked, he has been going to sea in his imagination, spinning out his tale of the relentless Captain Ahab of the *Pequod* chasing the white monster to the ends of the earth. Here, among these hills, he has found the inspiration to write the most ambitious American novel of the century, creating in Ahab a character to rank with the best of Shakespeare's tragic heroes, a wounded soul at war with the world and raging against it at every turn with curses hurled at man, beast, and God.

For the author of such a major work, Melville is still very young. At thirty-two, he is handsome in the rugged, masculine way of a young outdoorsman. Tall for the times, he is broad-shouldered and bearded, with dark brown hair that is thick and glossy, and blue eyes that are ever curious and alert. His own early adventures at sea on whaling vessels and an American warship are now well behind him. Eager to make his mark in the world, he has been trying to win fame as a writer almost from the moment his last ship docked, seven years ago. He has made remarkable progress, with several books now to his credit, each written at blazing speed, and most of them earning him praise if not a lot of money.

Published in November, *Moby-Dick* is far superior to anything he has done before. It raises its basic tale of a whaling voyage to the level of an epic adventure and a spiritual odyssey. This is supposed to be his breakthrough work, a potential bestseller that will establish him as an author with few peers. It has only recently landed on the shelves of the local bookstore in Pittsfield, and the response has not been good. Buyers have been few. In fact, the novel is selling poorly everywhere, and though there are several favorable reviews, the bad ones are especially damning. "Tiresome," "inartistic," and "not worth the money asked for it" are some of the milder criticisms in the American press. The worst attacks portray the author as a clever rascal determined to imperil the reader's soul by "piratical" assaults on "the most sacred associations of life." One critic is so outraged by the novel's impieties that he confidently prophesizes divine retribution as the price of the author's literary sins. "The Judgment Day," declares the reviewer, "will hold him liable for not turning his talents to better account."[3]

Even here in Pittsfield some of the criticism has been harsh. An old puritanical streak among the town's best families has caused them to shun the book. They have been shocked to hear the author condemned so forcefully for his irreverence. "The serious part of the community about here," Melville has learned, "have loudly spoken of the book[,] saying it is more than Blasphemous." Deeply in debt from the purchase of his farm—little more than a year ago, when he moved abruptly from New York—he has pinned all his hopes on his masterpiece paying rich returns. Now the grim fact is slowly beginning to emerge that his earnings will be paltry. Throughout the rest of his life, the American sales of *Moby-Dick* will bring him only $556.37 on sales of just over 3,000 copies. The book is an unmitigated commercial disaster.[4]

What should have been the happy close of a triumphant year, a time for celebrating the creation of a groundbreaking work, has instead become for Melville a sobering moment of public defeat. Sensitive to criticism, though often feigning indifference to it, he could be forgiven for avoiding any festive celebrations in the neighborhood and nursing his battered pride at home beside a warm fire. Yet here he is at the holiday party standing beside the most remarkable woman he knows, the new mistress of Broadhall. He seems to have some idea that a surprise is in store for him.

WHEN DINNER IS ANNOUNCED, he takes his hostess by the arm and leads her into the dining room, leaving her husband to follow, as if this is Melville's home, and Sarah is his wife. When they reach the table, "a beautiful Laurel wreath" lies before them on a plate gleaming in the candlelight, the handiwork of Mrs. Morewood, who has a talent for floral design. Without a word, she picks up the wreath and gently lifts it to Melville's brow, pressing close against him on her toes because he is so much taller. For a moment they look like actors playing a scene in an old drama. With a little imagination, this looks like the moment onstage when a queen crowns her champion or a maiden shows her favor to the victor of a race.[5]

At this crucial time in Melville's career, when his fortunes are sinking and the town is turning away from him, few gestures could carry greater meaning than Sarah Morewood's act of bestowing on him the laurels he deserves. In front of all her guests she is vividly demonstrating that at least one person in the community understands Melville's triumph, that *Moby-Dick* is not a "Blasphemous" failure, but a mighty work worthy of a crown. She knows even now what it will take the larger world several generations to discover—her neighbor has written one of the greatest novels in the English language.

And, in the spirit of solidarity, she has chosen this afternoon to join the author in a bit of blasphemy of her own, crowning a mere mortal on a day sacred to the Christian faithful. The modern mind may find nothing objectionable in her action, but what she does in this house at this time in a small New England town is shocking. Among Mr. Morewood's friends who share his devout Episcopalian faith, Sarah's tribute with her laurel wreath can't seem anything but a "pagan" custom that has no place on this day honoring the martyr who wore a crown of thorns. To the pious, the only proper ceremony for the dinner table will begin with the bowing of heads as Mr. Morewood leads his guests in prayer.

What Melville does next is neither pious nor predictable. He gives Sarah a graceful, though no less provocative, response. At the very moment that all eyes are on him, he declines the honor as gently as possible by lifting the wreath from his brow and placing it on Sarah's. Saying he will "not be crowned," he crowns her instead.

As his family will later note, Melville is "very angry" that his brilliant book has been damned by townspeople who are unlikely to have read it. Here in this highly theatrical episode at Christmas, with an audience of village notables, he and Sarah are staging a defiant ceremony that is both reckless and brave. It marks the beginning of a long period of deepening discontent for Melville, a grand but ultimately destructive turning away from worldly ambition and success.[6]

The dinner at Broadhall goes ahead, but not without what must have been a long and uncomfortable pause. No more is said of the wreath. It's put aside and remains on a table until the evening comes to a close. And then at the last minute—as Melville takes the reins of his sleigh, preparing to go—a servant comes forward, and gives it to him. Sarah won't let him leave without it. He rides away with the only prize he will ever receive for *Moby-Dick,* a simple token of honor from a woman who—within a generation—will be largely forgotten.

THE OBVIOUS, BUT UNSPOKEN, TRUTH HERE is that Mrs. Morewood is in love with Mr. Melville, who is also married. Indeed, Sarah will prove the most enduring influence on Melville's life, a muse as well as a lover. Yet the story of their affair has remained secret. This scene at Broadhall didn't even come to light until a century and a half after it took place. Details of the romance have been slow to emerge because much of the evidence was left unexamined while Melville's literary reputation languished in the late nineteenth century and early twentieth, and while Sarah's personal history faded into obscurity. Since then, biographers have shown little interest in a woman whom they have usually mentioned only in passing as a faceless "Berkshire neighbor." The most highly praised of Melville's recent biographers—Andrew Delbanco—devotes just four sentences to her in a book of four hundred pages. Elizabeth Hardwick's biography—the last on Melville in the twentieth century—doesn't mention Sarah at all, nor does Nathaniel Philbrick in his acclaimed book *Why Read Moby-Dick?,* nor do the sixteen distinguished scholars contributing essays on every aspect of the author's career in the recent Cambridge University Press *Companion to Herman Melville* (2014).[7]

The simple fact is that Melville's most passionate relationship—the powerful key to unlocking his secrets—has been missing from the story of his life. As a result, there has been more confusion and misunderstanding in his biographies than in those of the other great American writers of his time. What has been hidden is an affair so intimate and revealing that it colored every aspect of his life. It offers an almost modern insight into the pleasures and pains of sexual freedom.

The man who wrote *Moby-Dick* and filled it with such powerful, urgent longing on an epic scale, and then followed that immediately with a wild lament for forbidden love in the novel he called *Pierre*, didn't soar to such heights or plunge to such depths in an emotional vacuum. The tempests in those books had their parallels in his life, and at the center of the storm was a relationship for which he was willing to risk everything.

Herman and Sarah left a surprisingly long trail of clues behind them, some of it in letters, some of it in documents long buried in archives, and some in barely disguised revelations published in their lifetimes. Most of the information comes from the first few years of their affair, in the early 1850s, when they were most absorbed in each other's lives. This is also the greatest period of Melville's career, when he wrote not only *Moby-Dick* and *Pierre*, but also two masterpieces of short fiction, "Bartleby, the Scrivener" and "Benito Cereno."

Melville fell completely under his lover's spell from the moment they met in the summer of 1850. Mrs. Morewood was a singular character in the Berkshires of her day, a woman both bookish and beautiful, intelligent and inquisitive, creative and compassionate. Melville regarded her seriously as a kindred spirit, though his biographers have not. She is one of the great unsung figures in literary history. Yet her unconventional ways often made her the talk of Pittsfield, and the author of *Moby-Dick* was not the only neighbor who found her fascinating. A summer neighbor—Dr. Oliver Wendell Holmes Sr.—was so taken by her beauty and charm that he extolled them in verse, and wrote a novel largely inspired by her.

Even before she moved to Pittsfield, Sarah was known as an untamed spirit with a reputation for lavishing affection on her friends, male as well as female. Gossip was quick to spread wherever she went. By the end of her time in Pittsfield, she would have so many critics among the town's matronly guardians of virtue that one friend would say in her defense, "Her mistakes were nobler than some who criticized her; if she ever failed, it was thereafter to feel a more tender pity for the failing."[8]

The secrecy that attended her relationship with Melville was never impenetrable. Though relatively little of his correspondence survives, Sarah saved enough of his letters that we know beyond any doubt how he felt about her. Among the treasures in the public library of Pittsfield are several of these letters to her. They are the most romantic and lively of any he wrote in his life, and, taken together, they convey an overwhelming sense of the extraordinary release that he felt in her presence. Again and again, they portray her as a kind of mythical being who seems to be tempting him to achieve wonders in her honor. Though her position as an Englishman's wife living at a mansion provoked some of her neighbors to joke that Sarah was their Berkshire duchess, Melville was serious about exalting her. She becomes for him the reigning spirit of their neighborhood, "the peerless Lady of Broadhall," "My Lady Countess," and "Mrs. Morewood the goddess." And he becomes "Your Ladyship's Knight of the Hill." One letter ends with Melville playing to the hilt the part of a nobleman pledging his heart to a fair mistress in her castle. He signs off in an outburst of overcharged language: "With due obeisance & three times kissing of your Ladyship's hands, & salutes to all your Ladyship's household, I am, Dear Lady of Southmount, Your Ladyship's Knight of the Hill." (South Mountain—the highest point in Pittsfield—looks down on Broadhall.)[9]

AS HE DESCRIBES HER, she is not just a delight, but "Thou Lady of All Delight." She is not simply beautiful, but "the ever-excellent & beautiful Lady of Paradise." And, for Sarah, that description was not merely a fanciful tribute. The word *paradise* was loaded with special meaning, for it was a surname in her family. She was named after her mother, who was born Sarah Paradise. As the object of all this adoration, Mrs. Morewood returned the favor on that Christmas Day of 1851 in a manner only fitting for lovers who saw themselves acting out a courtly romance, crowning her "Knight of the Hill" with his well-earned wreath.[10]

Melville's passionate devotion to his "goddess" took him on a tumultuous ride from the grandest hopes of romance to a bleak, sobering reality. Understanding the great drama of this relationship is necessary to answer the most puzzling questions of the author's career. How did this young man known primarily for writing light books of adventure suddenly experience one of the most remarkable bursts of creative inspiration in literary history? And why was that short period in the 1850s followed by decades of relative silence and obscurity?

For about a dozen years Sarah and Herman lived next to each other in an isolated part of the Berkshires where they knew every mile. Set against so much natural beauty, their affair began in what was for them a kind of Eden. How they found their paradise, and how they lost it, is the story of this book.

PART I

—

A SAILOR IN THE BERKSHIRES

I love to sail forbidden seas, and land on barbarous coasts.

—*MOBY-DICK*

1

A SUMMER PLACE

The old mansion won his heart first, long before he met Sarah. He was just a boy on the verge of manhood, and in those days he could imagine that the house might one day be his when he was rich and successful. It is only one of the many strange twists in the story of Herman and Sarah that, decades before she acquired the place and named it Broadhall, the mansion and its 250 acres were the home of Melville's favorite uncle, his father's brother, who lived in it for more than twenty years as one of Pittsfield's most prominent citizens.

Young Herman was devoted to his uncle Thomas, whose warm heart and lively spirit touched him deeply. Looking back in later years to the times when he visited Thomas in summers, he recalled his uncle as "a cherished inmate" of his youth, "kindly and urbane— one to whom, for the manifestations of his heart, I owe unalloyed gratitude." The two would work alongside each other in the summer fields, raking hay in the warm sun. In their time together the boy came to know the land intimately, and the family joke was that his uncle's farm was Herman's "first love."[1]

Even then he thought it was a paradise. The fields sloped gently away from the house, with a patch of woods here and there full of maple and aspen, and then great clumps of wildflowers and raspberry bushes where the land flattened into a broad meadow. At the end of the property there was a wide view of the Housatonic River meandering through the valley, and high above on summer days the sky was a brilliant blue with occasional masses of white clouds drifting by.

The great jewel of the farm was its lake. Spring-fed and beautifully clear near the shore, it covered about five acres and was so deep in the middle that neighbors swore it was impossible to find the bottom. In the shallows there was a long bed of water lilies with blossoms in various colors depending on the time of year—pink, yellow, or white. For a while the lake was popularly known as the Lily Bowl. Visiting the Berkshires one summer, the poet Longfellow gazed at its shimmering surface and described it as the "Tear of Heaven."[2]

Uncle Thomas gave Herman not only an idyllic country escape, but also the inspiration to make his mark in the larger world. The old farmer had once been a man of great ambition himself, a young American banker in Paris, no less. Long ago, in the 1790s, Thomas had left his native Boston and spent almost twenty years trying to make his fortune in France. He won and lost great sums, met Lafayette, saw Napoleon, and discussed politics with James Monroe. His star seemed on the rise when he married a beautiful young woman who was born in Spain and brought up in France. She had connections in high places and wore her long dark hair in the stylish ringlets of Juliette Récamier, in whose glamorous circle she moved. But when his banking career collapsed in the last years of Napoleon's empire,

Uncle Thomas moved back home and settled his French family in the Berkshires. His wife didn't survive the move, dying in her early thirties in Pittsfield, her fashionable life in Paris far behind her.

Having studied her portrait after listening to his uncle's many recollections of his old life in France, young Herman was inclined to romanticize her death, and would later write that Thomas's "foreign wife paled and withered a transplanted flower." He never forgot the wistful look on his uncle's face when memories of that lost life of love and glory in Paris would now and then brighten the old man's imagination. In boyhood, at the kitchen hearth of the mansion, Herman used to watch with fascination as his uncle sat "just before early bed time, gazing into the embers, his face plainly expressing to a sympathetic observer, that his heart—thawed to the core under the influence of the genial flame—carried him far away over the ocean to the gay Boulevards."[3]

The house became a magnet in Herman's life, drawing him back even after he had seen so much more of the world. Isolated and provincial though the area was, his uncle's stories made it resound with echoes of grand endeavors "far away over the ocean." The place was always a reminder not only of his uncle's joy at diving deep into a foreign life and relishing every minute of it, but also of his sorrow at losing that life and the woman who had graced it until misfortune uprooted them from France.

THOUGH MELVILLE SPENT most of his early life in New York State—first in Manhattan, where he was born on August 1, 1819, and later in the Albany area—the Pittsfield farm was in many ways his true home. At an impressionable age its quiet beauty enchanted him, and in those long summer days spent at his uncle's side, his grandest dreams of worldly adventure and fame first took flight.

When Sarah Morewood lifted her wreath to his head in 1851, both of them were keenly aware of his long history in the house, and both understood why he would not wear that crown. The dreams born under its roof were in danger of disappearing, and Melville was afraid that he had already lost his best chance for success. If *Moby-Dick* couldn't take his career to the loftiest heights, what could?

It's still possible to follow Melville's footsteps into the very heart of that dramatic scene with Sarah, for the mansion remains standing in the twenty-first century. Changes and additions have been made at various times, but in the great hall its stately elegance still evokes a slower time when the loudest sounds were the ticking of a clock or the sound of boots on the stair. The double parlors haven't changed much. There is still a rambling grandeur to the maze of old rooms upstairs, and the distant view from the side porch is very much as it would have been long ago, with Mount Greylock—the highest point in Massachusetts—a rugged mass on the horizon.

That so much of the past remains is surprising given that the mansion is now the Country Club of Pittsfield, and the farm has been replaced by a golf course, but the employees haven't forgotten the mansion's connection to Melville, and at least one of the veterans is convinced that the place is haunted by various ghosts from its long history. On winter nights, when the course is covered in snow and most of the town is asleep, the house can indeed look ghostly in the moonlight.

Old Uncle Thomas is not among those who died there. Like that of many in Herman's family, his habit of squandering money was so bad that he was forced late in life to seek a new fortune far from New England. Toward the end of the 1830s he settled near the Mississippi River at Galena, Illinois, and died there a few years later without succeeding at anything he tried in his last home. His son Robert returned to Pittsfield and made a final effort to keep the treasured mansion in the family.

Around 1848 the Berkshire newspapers began featuring advertisements for a "private boarding house" to accommodate summer guests. "The rooms are very large," Robert boasted, "with many conveniences not usually found in ordinary boarding houses, and the situation is unrivalled either for the beauty of its scenery or the salubrity of the air." The new arrangement attracted a steady stream of guests—some of them famous, such as Henry Wadsworth Longfellow, who quickly spread the word to friends that he and his family had found a pretty hideaway in the Berkshires worth visiting. "We are lodged here better than our imaginations dared conceived," he wrote his friend Charles Sumner, the Boston lawyer and future United States senator. "It is a grand old mansion which you should not fail to sleep in." He loved the "echoing hall," the views from the wide porch, the elms and sycamores lining the drive, the "delightful deep shadows, and far-off gleams of sunshine flecking the landscape," and picnic lunches on the "pebbly shore" of the little lake. When the summer came to an end, the poet hated to go, though he found the place so peaceful that he had been unable to do any serious writing there. "Farewell the sleepy summer," he wrote on the last day of his stay.[4]

AS ROBERT AND HIS WIFE, Susan, soon discovered, their deluxe boardinghouse was expensive to run, and it would never make much money. But the guests kept coming, including in July 1850 the Morewoods of New York. As much as Longfellow enjoyed his stay, Sarah would enjoy hers even more, and by the end of the summer the whole place would belong to her. Unhappy with city life, she wanted to be free to ride and walk in a healthy environment where she could feel close to nature, and where she could do as she pleased. So she talked her husband into buying the mansion and 250 acres of farmland from Robert's family.

All her friends and family had heard some version of her frequent complaint, "A life in the City is very dull to me." She had grown up in a large Dutch family just across the Hudson in what was then a rural area of New Jersey. Shortly after marrying her in April 1845, Rowland Morewood tried to make the former Sarah Huyler comfortable in a pleasant house in Washington Heights, but even that leafy area of old Manhattan was too close to urban life for her tastes.[5]

She would always be restless and dreamy, a bright woman with endless curiosity searching for an elusive happiness. Born September 15, 1823, in what is now Passaic, New Jersey, she was the seventh of nine children. Close to her mother and sisters, she was on more distant terms with her three brothers and her father, who was a man of modest means. From an early age she was aware of being different from most people in her little Dutch community. She wanted something grander from the world and spent much of her time imagining in solitude an escape to a more romantic time and place. "I have been alone so many years of my life," she wrote at twenty-eight, "and have missed very much[,] which if known earlier and enjoyed[,] would have been of benefit to me now and I therefore have need to seize upon that which yields me happiness so long as I do not in so doing injure the feelings of those who understand and know me. Those who do not and will not [understand me] may act and feel as they like & judge me as they like too—"[6]

The key for Sarah was always to be understood, not judged. But, of course, the world prefers to judge first, and she came to take some satisfaction from shocking those who refused to understand her. Fearing she had wasted too much of her life waiting for better things to happen, she was eager to "seize" pleasures wherever she found them, and before they could vanish. Health problems also made her worry that her time might be short. A difficult pregnancy in late 1847, when she was twenty-four, left her so weak that she struggled to regain strength for months afterward. As she complained to a sister-in-law, she lost a great deal of weight after her son William was born. "I am very *slim* to use a yankee word," she wrote, "and all my dresses are too large for me."

What restored her health was her first long stay in the Berkshires, in 1849, and she couldn't wait to return. The air itself was like a tonic to her. The whole area, she said, was "as lovely as nature can make it." And because so many other interesting people were discovering its charms at the same time, she looked forward to sharing the experience in 1850 with more adventurous companions than her conventional husband.[7]

Her wide reading in romantic books had filled her mind with images of lovers being swept away by strong passions, and poor Rowland Morewood couldn't rise to those heights. His passions were limited to making money and serving his fellow Episcopalians. In New York he was not only a warden at his church but also its treasurer. Over time he would come to appreciate the Berkshires, but initially he went there simply to help his wife improve her health.

Yet, even on that first visit to the area, Sarah had wasted little time before commencing a summer romance. The man who caught her eye then was a handsome young lawyer whose family was politically well connected. He was Alexander Gardiner, clerk of the United States Circuit Court of New York. His sister, Julia, was one of the most famous women in America. Only five years earlier, at twenty-four, she had briefly served as the First Lady of the United States after marrying President John Tyler, who was more than twice her age.

When Tyler left the White House, he depended heavily on his brother-in-law for legal and financial advice. Alexander Gardiner was his appointed biographer and close friend, and everyone assumed that the young man would one day become an important political figure on the national stage—if he didn't allow scandal to ruin his career first. It was no secret that he had a weakness for liquor and late nights in the company of pretty women at wild parties. At one raucous affair in Washington, he boasted that the drink flowed so freely that "the floor drank as much champagne as the guests." Witty and outgoing, and in no hurry to marry, he "toyed with women as he played the stock market, acquiring and disencumbering himself of them as the situation demanded."[8]

In his brief fling with Sarah, the situation became so demanding that he fled from her in fear. Even in his worldly estimation, she was far too open and free with her affections to be "toyed with." One false step and he was sure that a major scandal would engulf him before he could escape the Berkshires. He was a public figure with a government job, and the last thing he wanted was a charge of adultery hanging around his neck.

He didn't worry that Sarah or her husband would make a fuss. What concerned him was that the proprietors of the boardinghouse in Pittsfield would spread damaging gossip. Servants had apparently told Melville's relations at Broadhall* of some compromising scene

---

* To avoid confusion, this name will be used for the Pittsfield mansion even though it wasn't formally known as Broadhall until 1851.

"respecting Mrs M[orewood] and myself," as Alexander Gardiner later put it. Worse, Melville's cousin Robert was already talking of it within his "circle" of friends and neighbors. If such gossip went beyond Pittsfield, there could be "very unfortunate consequences"—or so the young lawyer acknowledged to his brother in an anxious letter. "Very unfortunate" was a lawyer's polite language in the nineteenth century for a catastrophe. It suggests that what the servants saw was something much more revealing than a kiss or an embrace.

Guiltily, young Alexander confessed to his brother that he had been "very indiscreet and imprudent." Given his reputation for womanizing, he probably did little to discourage Sarah's attentions until they became a problem for him, but he pretended that the trouble was mostly Mrs. Morewood's fault. "I have apprehended sometimes that Mrs M's familiarity with me might lead to such a result and have done everything in my power to avoid it."[9]

Whoever was at fault, it was a dangerous game, but the remarkable thing is that neither Sarah nor Rowland seemed to understand how dangerous it could be. A powerful New York lawyer with a former president as his friend could ably defend himself against scandal. The Morewoods were small fish in comparison to sharks like Gardiner. Yet it was he who couldn't wait to clear out of Pittsfield as soon as he sensed real trouble.

There was always a disarming touch of innocence to Sarah's character. She didn't like to take no for an answer and resented the suggestion that she could ever do anything wrong. Once, when neighbors began gossiping about a clergyman who was spending too much time as a guest in the Morewood home, she wrote that she couldn't understand why anyone would doubt her good intentions. As always, she hated being judged by those who had no wish to adopt her larger point of view. "Slander," she remarked, "is an evil which no one can bear without great suffering let their innocence be ever so clear in their friends' minds."[10]

It is something of a miracle that Alexander Gardiner's hastily scrawled letter to his brother has survived. He lived only sixteen months after he left Sarah and the Berkshires, and his fondness for partying apparently hastened his death. After coming home drunk for three nights in a row, he fell ill and died of a ruptured appendix. His letter to his brother is now held in the neo-Gothic splendor of Yale's Sterling Memorial Library among a vast range of other documents from John Tyler's family. If it weren't for the Gardiner connection to an American president, the letter might well have disappeared long ago.

Domestic intrigues and romances were common in the fashionable boardinghouses of the time. Alexander Gardiner's own mother had warned him against them, urging caution in his dealings with ladies who happened to be sharing a roof with him for even a brief spell. "A very general and rather distant politeness is all that is necessary until you find them out," she had said of the ladies, "and then very likely you will wish to be still more distant."[11]

Undaunted by Gardiner's quick retreat, and the Pittsfield gossip, Sarah was thrilled to return to the Berkshires in 1850, and she had a foolproof method of avoiding more trouble from cousin Robert at Broadhall. With Rowland at her side, she would offer to buy him out and turn the place once again into a private home where she could come and go as she pleased. But it would take some time to complete the sale. Meanwhile, perhaps not by accident, she would find herself that summer living for more than a month under the same roof with the man whose emotional investment in Broadhall was beyond measure—Herman Melville.

2

"CORSET, SKIRTS, OR CRINOLINE"

In the years just before Herman met Sarah, one of the most celebrated characters in American literature was a nude Polynesian beauty. An English racehorse was named after her, a racing yacht carried her name on its hull, and a popular song was written about her ("The tender light of her blue eyes / Was mild and deep as moonlight skies"). Men dreamed about her, and imagined themselves locked in her embrace. She was the star of the most erotic scene in any major American book of the time, a vivid moment when she stood nude at the front of a canoe using her only garment as a sail and floating along like a South Seas Venus, free and unashamed. This was, of course, the lovely Fayaway in Melville's first book, *Typee*, where he coyly remarks of her classic figure, "We American sailors pride ourselves upon our straight clean spars, but a prettier little mast than Fayaway made was never shipped aboard of any craft."[1]

For most of his career, Melville was known first and foremost as the "American Robinson Crusoe," a daring castaway who had lived among cannibals, frolicked with native girls, and then returned to share his beguiling tales with a curious public. At twenty-one he had sailed from New England on a whaling ship, enduring eighteen months aboard until abandoning the vessel in the Marquesas Islands, on the other side of the world. For the next two years he was an Ishmael of sorts, a wanderer in search of adventure, slowly finding his way home after visiting Tahiti, Hawaii, and South America. Part fiction, part fact, *Typee* allowed him to embellish the most dramatic of his experiences among islanders whose exotic culture both frightened and delighted him. But for almost every admirer of *Typee*, what mattered most was the author's discovery of one island native in particular—Fayaway. Before his Whale caught on in the twentieth century, Melville's name was kept alive by his fond portrait of life with this slender young woman.

Men were jealous. "Enviable Herman!" a reviewer exclaimed in the usually understated literary pages of the London *Times,* adding, "A happier dog it is impossible to imagine than Herman in the Typee Valley." Even the cautious Nathaniel Hawthorne was aroused by what he called Melville's "voluptuously colored" descriptions of "native girls." And when Hawthorne's wife, Sophia, met the young author he charmed her—as he did many other women—with an air of exotic mystery suggesting forbidden pleasures. "I see Fayaway in his face," she said. She also liked to call him Mr. Omoo, after his second book on his island exploits, *Omoo: A Narrative of Adventures in the South Seas*. Others simply called him Typee.[2]

In 1846, when Melville was just twenty-six, the publication of *Typee* gave its author the same thing that Lord Byron achieved when

Childe Harold's Pilgrimage first appeared—overnight literary celebrity. Or, as Byron put it, "I woke up one morning and found myself famous." Melville's publisher in England had also been Byron's, and the connection wasn't lost on the young American writer. When he received his first written request for an autograph, he was amused at his sudden celebrity and replied to his admirer, "You remember some one woke one morning and found himself famous—And here am I, just come in from hoeing in the garden, writing autographs."[3]

It isn't difficult to understand why the book resonated with so many readers on both sides of the Atlantic. Sexy women and friendly cannibals will always draw a crowd. *Typee* aroused too much curiosity for the public to resist. It was the kind of book guaranteed to make the Victorians squirm in all the right ways. The islanders were humans like themselves, but different enough to be regarded as creatures of another world where sexual freedom wasn't unnatural, and eating your enemies was thrillingly repulsive.

Melville's "peep" into native life was just revealing enough to make his readers shudder, yet keep them reading. One moment he was giving the reader a glimpse of naked flesh, and the next just a peep into the cannibal's cauldron. "The slight glimpse sufficed," he wrote; "my eyes fell upon the disordered members of a human skeleton, the bones still fresh with moisture, and with particles of flesh clinging to them here and there!" In sexual matters he knew how to draw the veil just when his contemporaries might lose their nerve and look away. "The varied dances of the Marquesan girls are beautiful in the extreme," he wrote at the end of an early chapter, "but there is an abandoned voluptuousness in their character which I dare not attempt to describe."[4]

This kind of titillation drove some of his more enthusiastic female admirers to distraction. They wanted to share their own adventure with the dashing sailor. He soon became—in the words of a journalist—"one whose name often lingers now in terms of adulation upon many rosy lips." Through the post, an especially ardent female fan—a married Englishwoman in New York with the impressive name of Mrs. Ellen Astor Oxenham—pleaded with him, "Typee, you dear creature; I want to see you so amazingly." Among the many women who found Melville's adventures in his tropical paradise irresistible was America's most outspoken feminist, Margaret Fuller. In Horace Greeley's *New-York Tribune* she happily recommended *Typee* to her readers, saying that "Othello's hairbreadth 'scapes were nothing to those by this hero . . . and many a Desdemona might seriously incline her ear to the description of the lovely Fayaway."[5]

Some of his sympathetic critics were amazed at his ability to get away with so much questionable description without being censored. No one, said a writer in the *New York Daily News*, "dances the tight-rope of propriety better than Herman Melville. You are always expecting to see him fall off, but he never does. Some of his scenes with the nude nymphs of the Marquesas are so carelessly, and yet so tenderly told, that we trembled when we first read his *Typee*, but he went clean through the conventional hoop without damaging either himself or the circle."[6]

The predictable backlash against the book by moralistic critics only made it more popular with both sexes. There was something tempting about a story that provoked so many overwrought attacks. Melville was called "the shameless herald of his own wantonness," and was condemned for sharing so candidly his "voluptuous adventures." Horace Greeley was so confused by the stark contrast between the book's obvious literary merit and its occasionally racy language that he said it was both brilliant and "unmistakably defective if not positively diseased in moral tone."[7]

What was so dangerous about Fayaway was Melville's unapologetic celebration of her nudity in an age seemingly determined to bury the female figure under various layers of garments. She stood out in the imaginations of so many readers precisely because she was so refreshingly "devoid of corset, skirts, or crinoline," as a British poet put it. In that respect her "savage" freedom looked much more attractive than the grotesquely bundled bodies of so many women among Melville's supposedly civilized contemporaries. (In *Typee* the overdressed ladies parading through the great capitals of the world are described as "moving in whalebone corsets, like so many automatons.")[8]

This was the kind of unconventional thinking that would get the author into more and more trouble as his career progressed. In the early going, however, the pure delight in his portrait of a strange but fascinating culture won him more fans than it did detractors. He pulled off a trick that few writers of his time even attempted—he made erotic confession seem almost innocent, as in his surprisingly matter-of-fact remark in *Typee*: "Bathing in company with troops of girls formed one of my chief amusements." With every subsequent book there was always a general sigh of regret by his most devoted admirers that he didn't give them another character like Fayaway.[9]

Whether or not every word of *Typee* was true never mattered much to those who loved its story. The book offered a glimpse of a paradise that was radically different from the grim realities of daily life in societies where too many natural freedoms were forbidden or harshly punished. For a young woman like Sarah Morewood, romantic visions of sexual freedom like those in *Typee* were the stuff of her dreams. She was one of those Desdemonas that Margaret Fuller imagined sitting in their parlors yearning to hear more from Melville.

THOUGH NONE OF HIS SUBSEQUENT BOOKS would enjoy the success of his first, Melville quickly turned out four more in the next four years. The reading public could hardly keep up as the young author gave them variations on his adventures at sea, tales that always seemed to mix fact and fiction, and to add further mystery and intrigue to the life of the exotic voyager. All of his titles sounded odd to his contemporaries, and none seemed self-explanatory—*Omoo* in 1847, *Mardi* and *Redburn* in 1849, and *White-Jacket* the following year. He wrote the last two at such astonishing speed that he needed only four months to complete both.

Not a few reviewers wondered if the same man could be the author of all these works. Before they accepted that the stories were based on the experiences of a single young man, some critics asked how anyone at that age could write so well and show such courage at sea and in remote lands. It was difficult to believe that Melville the Hero was not only a Pacific castaway, but also the endearing innocent sailing the Atlantic for the first time in *Redburn,* and the slightly older but more cynical navy man on the warship *United States* (or *"Neversink"*) in *White-Jacket*. Though a large part of all these stories was indeed drawn from his own life, British reviewers were especially dubious, asking whether they were being fooled by episodes created out of nothing by one or more writers hiding behind a false name. Protested one critic, "Herman Melville sounds to us vastly like the harmonious and carefully selected appellation of an imaginary hero of romance."[10]

At the height of his celebrity in the late 1840s, the author was such an object of curiosity that even his youthful past in the Berkshires became widely known. During Longfellow's stay at Broadhall when it was a boardinghouse, he jokingly gave his address as "Melville Hall, Typee Valley Pittsfield." On her visit in 1849 Sarah couldn't have avoided hearing about the area's newest celebrity—Mr. Herman Typee himself—and the fact that the old mansion on the outskirts of Pittsfield was a place that he occasionally graced with his presence.[11]

If Mrs. Morewood wanted to meet the author, then becoming the owner of Broadhall was an extravagant but effective way of drawing him to her door. For a woman desperately wanting a summer romance with a companion who might be steadier than the capricious Alexander Gardiner, Melville must have seemed an immensely attractive choice. The man who could lean back and calmly study Fayaway's fine features when she had removed her last garment was just the sort of fellow Sarah Morewood was hoping to find. The fact that another woman had already found him and married him was inconvenient, but it wouldn't stop her.

3

THE JUDGE'S DAUGHTER

In the heart of Boston's most exclusive district, at the very top of Beacon Hill, is a large townhouse with an enormous arched doorway of heavy stone that looks like an entrance to a jail or a bank vault. In a neighborhood that Henry James called "the solid *seat* of everything," this house—Number 49—seems as if it began life as a fortress and only gradually came to resemble the other comfortable but very solid homes on Mount Vernon Street. For two hundred years the rich and powerful have been affirming their success by buying houses in this long thoroughfare that begins near the Charles River and rises all the way to the back of the State House. In Melville's time there was no more powerful resident of this street than the man who owned Number 49, Lemuel Shaw, the chief justice of the Massachusetts Supreme Judicial Court.[1]

Shaw's many friends regarded him as the greatest man in Boston. A future United States senator from Massachusetts would recall that Judge Shaw was "venerated as if he were a demi-god." People saw him as a model of judicial integrity and fairness, a solid defender of the law with a gruff, no-nonsense manner but a soft heart when compassion demanded it. He had no patience for insincerity in his courtroom and wouldn't hesitate to use his booming voice to silence upstart lawyers trying to sway him with theatrics. When an officious attorney waved a book in front of him, crying, "Look at the statutes, Your Honor, look at the statutes," the judge thundered, "Look at them yourself, sir."[2]

When he was offended, Judge Shaw made an expression not unlike that of an angry bloodhound. Heavy and broad-shouldered, he had a big, jowly face with a prominent nose and shaggy brow. With his spectacles balanced on the tip of his nose, he could glower at an impertinent lawyer and stop him midsentence with a low rumbling of displeasure. One day, an attorney who had frequently been admonished by Shaw was seen walking a large dog on a leash and was asked where he was going. "Down to the Supreme Court," he replied. "I thought I would show him the Chief Justice so as to teach him to growl."[3]

At his home in his spacious brick townhouse, this mighty bulwark of justice could act more like a sleepy, harmless spaniel. An avid reader with a taste for history, he was most at ease in the warm comfort of his second-floor study, with long shelves of books on every wall. He liked the eighteenth century—he was born in 1781—and was fond of his Hogarth engravings and of the evenings he spent playing whist, the fashionable card game of his youth. His second wife—Hope—doted on him and even put aside her religious scruples to play cards with him late at night before Sunday services the next morning. When he had finished with the serious work of the day, his voice rang through the house as he playfully commanded his wife to join him for cards. "Hope, come here and have a game," he would shout.[4]

He was especially indulgent toward the only daughter among his four children, Elizabeth. There was a son by his first wife, who had died giving birth to Lizzie—as she was known in the family—and two more sons by his second wife. He was always generous to his daughter. When he gave a coming-out party for her, the very best of Boston society showed up—two hundred in all—and he spared no expense for the proud occasion. He hired musicians, took up the carpets for dancing, and spent lavishly on the food and drink. There was champagne, claret, and sherry, with ham, pâté, and oysters served three ways (scalloped, stewed, and fried), and then cakes, ices, and truffles for dessert. He could afford such extravagance. Before becoming chief justice in 1830, he had been one of New England's most successful lawyers, and had amassed a tidy fortune of one hundred thousand dollars.

Because of his wealth, his daughter should have had her pick of Boston's most eligible bachelors, but she was a quiet, unassuming young woman who seemed happy living in her father's large shadow. Refined and dutiful, with a plain face and a prominent nose too much like her father's, she didn't have a reputation for turning heads. She seemed most likely to marry a hardworking but unexceptional lawyer or schoolteacher.

Given his background as a largely self-educated man who had become famous for paddling around a Pacific island with a nude "savage," Melville wasn't the logical choice for Judge Shaw's daughter. But marry her he did in August 1847, at the height of his new fame, and with her father's fond approval. She was a far cry from the dark romantic maiden that readers of *Typee* might have expected their author to wed.

FANNY APPLETON LONGFELLOW—a great beauty of Beacon Hill who visited the Berkshires with her husband, the poet—was astounded when she heard of Melville's marriage. She had read *Typee*, and as a former neighbor of the Shaw family she knew all about the judge and his daughter. In private she remarked of Herman and Lizzie, "After his flirtations with South Sea beauties it is a peculiar choice (in her)." No doubt many others thought the same. In temperament and intellect, the couple had little in common. Elizabeth was uncomplicated, practical, and straightforward, with few interests beyond friends and family. She wasn't artistic or literary, didn't seem to care much for travel, and rarely stood out in a crowd.[5]

For those who knew little about the bride except her name, the match seemed a fairy tale in the making. Far away on the shores of Lake Michigan, the *Chicago Tribune* took the news of the marriage as the final proof that the celebrated writer was leading a charmed life. "Five years ago Herman Melville was a sailor before the mast in a South Sea whaler, a fugitive, and a prisoner. Now he is a famous author, 'the Phoenix of modern voyagers,' and has just been married to a daughter of the Chief Justice of the State of Massachusetts. The novelist never imagined a series of more romantic adventures than these events of real life."[6]

The truth is that the marriage came about largely because the Shaw and Melville families had been connected for many years. In his youth the judge had been in love with one of Melville's aunts. She had died before they could marry, but he had remained a friend of the family ever since, and had been especially close to Herman's father, Allan. Like his brother Thomas, Allan was good at losing money faster than he could make it, and he sometimes turned to his friend Lemuel Shaw for legal advice and other assistance. But no one could save Allan from financial disaster.

Urbane, kindhearted, and ambitious, Allan had once been a prosperous merchant in New York, importing goods from France, where he traveled widely. He gave his wife, Maria Gansevoort, and their large family—which grew to include four boys and four girls—a comfortable life in a series of well-furnished Manhattan homes. But with merciless speed his business collapsed under too much debt, and he was forced into bankruptcy. Herman was only ten when his father's troubles began.

Allan tried to revive his fortunes in Albany, his wife's hometown, but he continued struggling to pay his bills. Two years later— broken and humiliated—he died after a short illness. He was only forty-nine. The family fell on hard times, Herman left school to work in an Albany bank, and life was never the same. The proud, determined Maria, whose ancestors were Dutch gentry, settled into a long widowhood and did her best to care for her family. "Oh the loneliness," she would later say, "the emptiness of this world when a woman has buried the husband of her youth & is left alone to bring up their children."[7]

A domineering, self-righteous figure in her family, Maria expected her children and other relatives to provide for her. She used her religious faith to instill guilt not only in her children, but also in the heart of her family's most generous benefactor—Judge Shaw. Pleas to him for help were accompanied by Maria's reminders that "the sincere prayers of the Widow & Children shall ascend for your repose here & hereafter." Herman grew up regarding the judge as something of a father figure, so Lizzie Shaw was almost like a cousin, and the two had met long ago.[8]

It is impossible to gauge the depth of Melville's feelings for his bride. No letters survive that would provide some glimpse into his heart. He and his family lost or destroyed so many of his letters that only one survives from him to the woman who was his wife for more than forty years, and it is only a routine item of little interest. Any trace of the dreamy, romantic side of the young man who had run off to sea is hard to find in the author who agreed to wed the judge's daughter. In settling for the unspectacular Lizzie as his wife, the promising new writer was apparently willing to forgo an American Fayaway for the security of a generous father-in-law with influence and high standing in society.

He had already used his connection with the judge to promote *Typee*. Anticipating that many critics would question the more sensational events in his first book, he drew attention to his relationship with the great man of the law by dedicating the book to him, hoping that the older man's sterling reputation would lend his tale some credibility. Following the title page, this prominent tribute appeared in the American edition of *Typee*: "To LEMUEL SHAW, Chief Justice of the Commonwealth of Massachusetts, this little work is gratefully inscribed by the author." (The British edition substituted "affectionately" for "gratefully.") The judge was flattered, and presumably Melville penciled this in with Judge Shaw's permission. Because Shaw cared about language, loved books, and had been kind to the Melville family, it made sense for Herman to seek a little protection behind the judge's very large shield of integrity. No doubt many readers were impressed, but some of the book's critics were shocked. "It is a matter of surprise to us," declared the *Christian Parlor Magazine*, "that such a work could have obtained the name of LEMUEL SHAW."[9]

The judge must have had a few misgivings about handing his daughter over to a man with no steady job and a vagabond past. If he had been more objective, Shaw might have used his sharp legal mind to question how a pampered young lady of Beacon Hill with ordinary looks and no special talents could please the extraordinary hero of *Typee*. Lizzie seems to have had her own doubts, because one of her great fears before the marriage was that the church would be overrun by all the envious women who yearned to be Melville's Fayaway. In the end she insisted that the wedding take place at her home in Mount Vernon Street instead of at the family's church.

They were married in the late morning of August 4, 1847. Lizzie was twenty-five and Melville had just turned twenty-eight. What would endure in the bride's memory was "a vision of Herman by my side, a confused crowd of rustling dresses, a row of boots, and [the Reverend] Mr. Young in full canonicals standing before me, giving utterance to the solemn words of obligation."[10]

AS A FIRST TOKEN of his generosity to the couple, Judge Shaw helped them to acquire a twenty-one-year leasehold on a house in Manhattan large enough to accommodate not only the newlyweds but also Melville's mother, a younger brother and his wife, and—as if that weren't enough—all four of Herman's grown, but unmarried sisters. It was a good first step in putting his mother and siblings back where they had started as a respectable family living well in New York in earlier times, and he couldn't have done it without the judge's help.

The house was in a decent neighborhood on Fourth Avenue behind the recently constructed Grace Church. The Melvilles would be one big family under one roof—as though Lizzie had always been among them, like an adopted daughter. Most newlyweds would have sought a more private love nest than this crowded house, but Lizzie acquiesced to the arrangement and wrote reassuringly to her family in Boston that she was enjoying her new life, though she had never been to New York before. "I'm afraid no place will ever seem to me like dear old crooked Boston," she told them, "but with Herman with me always, I can be happy and contented anywhere."[11]

At the outset, there was certainly affection and warmth in this relationship, but no sign of any great passion. As Melville would lament in one of his later poems, "few matching halves here meet and mate." One of his cousins thought the feelings between the couple were more "ethereal" than physical. Many decades later, after both were gone, the couple's granddaughter Eleanor Metcalf would tell an early biographer, "You say, in your *Nation* article, that Melville was happily married. He wasn't."[12]

For three years Herman did his best to make it all work, to keep Lizzie and the rest of his family happy. In less than two years he became a father. His son Malcolm was born in February 1849. Proud of this new addition to the family line, he gave him a name that paid tribute not to his mother's Dutch background, but to his father's Scottish roots. He boasted that he was "of noble lineage—of the Lords of Melville & Leven." And, in fact, in Elizabethan times one of Herman's ancestors held a knighthood and a small castle by the edge of the sea near Edinburgh. (In America the family had been spelling the name without the final *e,* but after the death of Herman's father his class-conscious mother decided a change was needed. So the widow and her brood abandoned "Melvill" for what was then considered the more distinguished "Melville.")[13]

While Lizzie cared for her new infant, her husband scribbled away upstairs. His oldest sister—Helen Maria—helped out by making fair copies of his manuscripts, keeping pace with her brother page by page. Grimly diligent, he went to his desk each day as though to a barn to do chores. Needing to produce something sufficiently commercial, he reported to Judge Shaw that he felt compelled to keep writing "as other men are to sawing wood."[14]

Then, after a brief time abroad in England at the end of 1849, he began trying to write a different kind of story—something that might please both the reading public and himself. His literary efforts weren't earning him enough money to support his large household, and he needed a breakthrough book that would not only match *Typee*'s success but also reach an even larger audience. He had made moves in this direction with *Mardi*, but its added layers of allegory, philosophy, satire, and social criticism make an uneven fit with its basic tale of the sea, and the book did not do well. By July he was well into a story that drew on his old whaling experiences, a big book in theme and scope about a nautical world that he was uniquely qualified to explain. It was still missing something "to cook the thing up," as Melville put it. "One must needs throw in a little fancy."[15]

In the next year "a little" would turn into a lot, with his novel of the Whale increasingly elevated through "fancy" into a more lyrical and mythic tale. In his daily life, however, Melville needed a change of scenery first. As the temperatures rose that summer, he didn't find it easy working in New York. With so many people living in his house, the noise and heat became increasingly hard to bear. The weather turned so bad that the newspapers warned of the city reaching its "melting point." There was no relief from the burning sun, the *Tribune* observed, "unless one was able to sit all day eating ice-creams, with his feet in a tub of water, and the lightest possible clothing on his back." Before the city became too oppressive Herman Melville was gone. One day in the middle of July 1850 he packed a bag and left on his own for what he thought would be a short escape to the cool heights of the Berkshires.[16]

4

THE FIRST STEP

It was early in the summer of 1850 that Sarah Morewood arrived in Pittsfield for her second season in the Berkshires. Rowland accompanied her there from New York by train, and not long after he had negotiated his purchase of Broadhall, he returned to the city, coming back for weekends when his work allowed. It was a pattern that he would follow for years, and Sarah would dutifully tell his relatives in England how much she regretted "that my husband is with me so little." Broadhall and the Berkshires were her playground, and Rowland acknowledged that fact by staying out of the way as much as possible. For the sake of his family's business in the city, he couldn't do otherwise. The firm was called George B. Morewood & Company, and it was located in lower Manhattan, where it specialized in selling galvanized spikes, bolts, wires, and lightning rods. Nothing could be less poetic, and Sarah rarely mentioned the business. Her own parents had little money, so every penny she spent in her beloved Berkshires, Rowland had to earn in a brick warehouse at 14 Beaver Street.[1]

Making Sarah happy seemed to make Rowland happy. She was the only touch of magic in his otherwise routine life. He wasn't handsome, just wealthy. His face was round and plain, with small, soft eyes that seemed to disappear in the back of his head under a heavy brow. In later years he grew stout, his beard turned white, and he lost most of his hair. There was almost nothing he wouldn't do for her. Broadhall didn't have a piano, so he had one shipped from Albany for Sarah to use just for the summer. If she needed more carriages to take friends to a picnic, she hired whatever was required in town. Though Broadhall continued to operate as a boardinghouse for the rest of the season, Sarah acted as if the mansion belonged to her even before the sale was finalized. She was treated as the owner in all but name and brought her own servants for the summer. (Whenever Rowland was at the house that year, he gave his address as "J.R. Morewood, At the Melvill Place, Pittsfield.")[2]

It may be that Melville's cousin was desperate to be rid of the costly mansion, or that Rowland was too clever for him, but the purchase price was a steal. A similar farm in the neighborhood sold a few years later for $18,500, and the house on that land was not nearly as impressive as Broadhall. Yet the Morewoods were able to acquire their magnificent summer place for the absurdly low price of $6,500. Rowland had the advantage of being able to pay cash, and because the deal was sealed in private, without any notice to the general public that the place was for sale, there were no competing offers to consider. The rock-bottom price must have delighted Sarah, who wanted to make some improvements. Whatever money was saved on the purchase would be used later that year for renovating the kitchen and other rooms.[3]

When Herman Melville happened to show up at the old familiar mansion in mid-July 1850 to visit his cousin, he walked into a house that may have looked the same but was suddenly filled with a fresh air of excitement and purpose. Sarah was full of big ideas for her future in the Berkshires, and was busy inviting various members of her New Jersey family to come and see her new house. She was eager to throw large parties, organize elaborate picnics and dinners, launch extensive tours of the region, and make every summer at Broadhall unforgettable.

The Berkshire landscape was the perfect backdrop for her. It seemed to draw out her good looks and bring her best qualities to the forefront. Her dark, enigmatic face was all the more seductive against the background of a natural world so rich and colorful in summer and autumn. One of her best friends—the journalist Caroline Whitmarsh—said that Sarah "gloried in the beautiful scenery" of the region and eventually became one of its most knowledgeable guides. Any bright spot in the landscape would arrest her attention and produce a cry of pleasure. "She was never so happy as when pointing out [its] beauty to others," wrote Whitmarsh of Sarah and the Berkshires, "dipping her hands in the stream, bending over the flowers, looking up through leaves at the sky, or dreaming across the lake, with a true lover's love."[4]

There is no way to know how much Melville had learned about Mrs. Morewood before he arrived, or how much her plans depended on meeting him. Once he returned to this Berkshire version of Typee Valley, he slowed work on the whaling book while Sarah captured more and more of his attention. If he resented her ownership of Broadhall or worried over her character because of anything his cousin Robert might have told him, it never showed.[5]

EVER SINCE HIS RETURN FROM THE PACIFIC, Melville had been leading a relatively quiet life. Unlike Alexander Gardiner, he had no reputation as a man who liked to party or chase women in the big city. Only his books told of his more liberated and indulgent self. Family life seemed to tame him, but on the sea or among the maidens of the Pacific, there had been a different Melville who relished adventure, pleasure, and danger—and boasted of it. Under Sarah's influence his bolder self emerged once again. He returned to enjoy a summer devoted to activities that he had rarely or never undertaken in mixed company before—champagne picnics in green meadows, costume parties, late-night dinners, and long, lazy afternoons boating or fishing at a wooded lake.

Lizzie Melville—and others from her New York household—came to Broadhall that summer to join Herman, but she and her in-laws are rarely mentioned in the various accounts of that period by visitors and neighbors, and seem to fade into the background. As an escape from New York, what Lizzie enjoyed most was the safe, predictable life at her father's home on Beacon Hill. In Manhattan she had made little impression on anyone, and was content to avoid the literary spotlight that was sometimes turned on her husband. She was convinced that he was now essentially a homebody, too. "Herman is not fond of parties," she had declared from their new place in New York soon after their marriage, "and I don't care anything about them here."[6]

When Lizzie arrived in the Berkshires for the family holiday, she was in for a surprise. Suddenly Herman couldn't resist parties and pleasure excursions—if Sarah had anything to do with them. Mrs. Morewood was the center of his attention, and he showed no interest in returning to New York that summer as long as she was at Broadhall.

When two of his city friends—the editor Evert Duyckinck and the poet Cornelius Mathews—visited him for a few days, the one woman who stood out for both of them was Sarah. They barely noticed Lizzie. To Mathews, Mrs. Morewood was indisputably "the sorceress of the scene," dominating and directing every activity wherever she appeared. Duyckinck was similarly impressed. Watching her crack a whip as she drove a pair of horses down a country lane, he remarked with sudden wonder, "There's a woman with a snapper." To his wife in New York, he commented that Sarah would soon be the "owner" of Broadhall, and he promised, "I must tell you more about her."[7]

On a long carriage excursion one afternoon to Pontoosuc—a large lake on the other end of town—Melville first experienced Sarah's ability to command attention and shine against the Berkshire backdrop, and he quickly fell in with her festive plans. In a procession of carriages, Sarah treated Melville and his friends to a lively tour of the surrounding countryside, springing up in her seat every so often to stand on tiptoe and point out various natural wonders "hid away in the distance." Cornelius Mathews was so charmed by her spirited commentary and shapely figure that he quickly adopted the view of so many of her admirers—that she couldn't be entirely human. Soon he was babbling away about mountain fairies in storybooks, and how Mrs. Morewood must be an actual fairy sent down to weave a spell around visitors like him. (Fanciful but homely, Mathews never had much luck with women, and even his male colleagues tended to mock his looks. James Russell Lowell called him "a small man in glasses.") Even among old-time residents Sarah was often the first to notice an overlooked area of beauty. Perhaps as a result of his rambles with her, Melville remarked in August, "It is curious, how a man may travel along a country road, and yet miss the grandest or sweetest of prospects, by reason of an intervening hedge, so like all other hedges, as in no way to hint of the wide land-scape beyond."[8]

In the Berkshires that summer it seemed that Sarah Morewood was everywhere. She was up early and often dashed across the horizon on horseback or at the front of a carriage with a cloud of dust trailing behind her. An accomplished rider, she thought of her horses as friends and loved the freedom they gave her to go anywhere she wanted at a moment's notice. By the end of the season, one of Melville's vivid memories would be of the "sprightly" Sarah on the back of a horse, "patting his neck & lovingly talking to him." Among her horses in later years, there was "a fine filly" that she named—of course—Fayaway.[9]

CORNELIUS MATHEWS THOUGHT that his stay with Melville in the Berkshires was like a voyage to a romantic land, and it would appear that Sarah did much to create that impression for him. After his holiday at Broadhall was over, he declared himself one of her devoted well-wishers when he inscribed his latest book to her, and he explained to the readers of Duyckinck's *Literary World* that he had spent part of his summer "sailing" in the Berkshires: "We launched out, on our first entrance into the new region, like mariners upon an unknown ocean, ready to make the most of every current and islet on our course."[10]

Melville had a similar feeling. Writing two years later in the novel heavily influenced by his Berkshire experiences—*Pierre; or, The Ambiguities*—he rhapsodizes over the countryside with the same passion and fond attention to detail that he gives to the sea in his other books, finding that the dark forests recall the beginnings of time and nature's supremacy in the same way that "the eternal ocean" does. Just as his impending voyage on a whaling ship awakens Ishmael in *Moby-Dick* to the majesty of the natural world ("the great flood-gates of the wonder-world swung open"), so the young hero of *Pierre* surveys a landscape like that surrounding Broadhall and asks hopefully, "Is it possible, after all, that . . . this world we live in is brimmed with wonders, and I and all mankind, beneath our garbs of commonplaceness, conceal enigmas that the stars themselves, and perhaps the highest seraphim can not resolve?"[11]

It is a young woman with long, dark hair and penetrating eyes—a beauty of "intense and fearful love for him"—who inspires Pierre to see the natural world with fresh eyes. He concludes that she has emerged from "the wonder-world" to show him a new life full of mystery and magic. He is one kind of man before he meets her, and then she turns his world upside down, shattering his previous sense of dull routine and turning him into an impassioned creature barely recognizable to those who knew him before. Powerless to resist her, he describes her as "so beautiful, so mystical, so bewilderingly alluring."[12]

A quick look, a touch of her hand, her upturned face in a doorway, her musical voice, a vision of her by lamplight—in every aspect of her appearance and character Pierre finds something to stir his heart, and he soon believes that he has discovered at last the kind of love he has only previously known in books. He imagines that he and the young woman are like two figures in a courtly romance. Vowing to be her knight, her champion, he regards her as his Lady, and addresses her in the same archaic, stylized language of all those courtly letters that Sarah would save from Melville.

Because the author spent the first weeks of his Berkshire holiday under the same roof with Sarah, they probably didn't exchange letters while they stayed at Broadhall. But, as a future chapter will show, *Pierre* is so close to the known facts of their lives that it can reveal much of the behind-the-scenes drama. The novel is a love letter both to the woman who changed his life and to the world of Broadhall that served as an inspiration to each of them.

AS A COMPANION PIECE TO *MOBY-DICK*—the backstory to the creation of the whaling saga—*Pierre* will be easier to understand later in the story of Herman and Sarah. For now, however, there is one clue worth considering that has managed to withstand the passage of time, yet can still speak eloquently to the emotions of the couple's early days. For the past half century this clue has sat on a shelf in a redbrick townhouse on a quiet street in an old neighborhood of Philadelphia. On the third floor of the little Rosenbach Museum and Library, there is an expensively bound volume published in 1854 with the following words written inside in black ink: "H. Melville Pittsfield Mass. Presented to Mrs JR. Morewood." There is no note to indicate when he gave it to her, but the time was probably toward the end of the 1850s, when illnesses and other setbacks had cast a shadow over their lives. Melville, it seems, wanted to recall better days with an apt quotation hidden in plain sight in the pages of this book.[13]

Simply as an object, *The Poetical Works of John Dryden*—a Victorian collection of the Restoration poet's work—is exquisite, with crisp, gilded pages, beautiful illustrations, and sharp print. Back and forth, over the years, as Herman and Sarah would exchange such gifts, some of those around them seem to have assumed that they were merely two book lovers sharing a passion for the written word in fine editions meant to last. But, as they were quick to realize, books are wonderful places for hiding secrets.

Dryden wasn't really Melville's kind of poet, nor Sarah's, and the pages seem relatively untouched, especially in the first half of the book, where most of the famous poems appear. In the very back, with its less familiar titles, a few things stand out in an otherwise smooth sea of white margins. Melville, who had a lifelong habit of leaving penciled marks in his books whenever a line or passage caught his interest, singled out some passages for Sarah's attention. The most noticeable one occurs in Dryden's highly erotic "Sigismonda and Guiscardo," an adaptation of Boccaccio's tale of a secret love affair between a princess and a young man at court. Closely watched by her possessive father, Princess Sigismonda must conceal her passion and communicate with her lover through hidden messages and furtive looks. Desperate to be alone with Guiscardo, she arranges a meeting in a secret cavern where they make love with such violent intensity that they forget everything in the outside world. " 'Twas restless rage, and tempest all the night," writes Dryden. "Love rioted secure."

With a check mark in the margin, Melville highlighted the passage where the eager young man announces his arrival at the cavern for his night of pleasure, and the princess races forward to embrace him. In one of the most romantic lines of the poem, Dryden writes of the young man, "And the first step he made was in her arms." It is specifically this line that Melville checked.

No married man in the 1850s would have pointed out to another man's wife a poem as obscure, and as erotically charged, as Dryden's tale unless he knew she would welcome it, and understand its significance. But this was part of a fanciful, romantic world that both Herman and Sarah found appealing, this dreamy realm of lovesick heroes and heroines. A few lines, most of them faint, appear in two other romantic poems by Dryden—both dealing with love at first sight, and both also near the back of the volume. Some of these lines look as if they were erased, but overall what the marks seem to suggest is that Melville was searching for just the right passages to share with Sarah. If the passage with the check beside it is intended to serve as a reminder of their earliest encounters, its larger context is certainly in keeping with Melville's celebration of her in his letters as "Thou Lady of All Delight":

He came, and knocking thrice without delay,
The longing lady heard, and turn'd the key;
At once invaded him with all her charms,
And the first step he made was in her arms.

5

THE YOUNG TURK

Young Herman Melville liked showing off, and as his first summer with Sarah progressed, he found a promising opportunity to prove how daring he could be in the right circumstances. It was on a Friday in August 1850, when Sarah was excitedly preparing for her first costume party at Broadhall.

The morning began with an exuberant search for old clothing to create costumes for the evening. Sarah led a group of friends through the cellar and the barn to dig out big hats, boots, and fancy jackets from storage, relics of the days when Uncle Thomas was the town dandy. In anticipation of becoming the owner of a late-eighteenth-century mansion, she wanted to appear at her party as an extravagant lady of the era, with all the fashionable trimmings she could find for her outfit.

The first shock of the day was her decision to dress not as a typical aristocratic socialite, but as one of the bawdiest characters in eighteenth-century fiction—Aunt Tabitha from Tobias Smollett's picaresque novel *The Expedition of Humphry Clinker*. Sex-obsessed, and incorrigibly flirtatious, Tabitha was hailed by Sir Walter Scott as one of Smollett's most memorable characters, but she was notorious for her carelessly obscene language. Smollett portrays her as a wealthy, middle-aged woman who has spent her whole life chasing men with little result. "She has left no stone unturned to avoid the reproachful epithet of old maid," says her nephew. She is determined to find a husband who will not only satisfy her ravenous appetite for love, but will also tolerate her habit of writing so hastily that she often litters her correspondence with unintentionally hilarious sexual puns. She says that a man "beshits me on his bended knees" (she means beseeched); demands that a mattress be "well haired"; and tells her female servants that her "fervent prayer" is to see the family's new footman—Humphry Clinker—"penetrate and instill his goodness, even into your most inward parts." She sends a reminder home "to have the gate shit every evening," and complains that troubles with a man named Roger have caused her to be "rogered." Perhaps the most infamous of her malapropisms is her written reminder to her household "to keep accunt of . . . the buttermilk."[1]

This eighteenth-century humor was still part of the comic banter of sailors in the nineteenth century, and Melville had heard it all in his years at sea. (Smollett himself was a naval man, having served on a British warship as a surgeon's second mate, and later achieving his first literary success with a novel that drew heavily on his seafaring experiences—*The Adventures of Roderick Random*.) The surprising and arousing novelty here, in the well-heeled Berkshires, was that Melville had found a pretty woman who could enjoy a good joke like a sailor. At least one of his literary friends from New York—the ever curious Cornelius Mathews—recognized, or was told, that Sarah was dressing as Aunt Tabitha, but it's doubtful that anyone in Melville's family would have known enough about *Humphry Clinker* to understand the meaning of her choice.[2]

When Melville himself announced that he was coming to the party as a Turk, even Mathews may not have realized that the young author was matching Sarah's provocative costume with one of his own from Smollett's racy tale. A scene full of sexual innuendo takes place in the novel when a bearded Turk in full-flowing traditional dress, complete with turban and scimitar, enters an English drawing room and is overwhelmed by a wave of cultural misunderstanding. The great illustrator Thomas Rowlandson made the scene famous in a print depicting the clash between the tall, dignified Turk and an imbecilic English duke. So ridiculous is the conversation of the English aristocrat that the Turk decides the whole country must be "governed by the counsel of idiots."

Even at this early stage of their relationship, Sarah and Herman were reveling in their hidden, bookish jokes. As John Updike has pointed out, Melville's "sense of truth held him stubbornly close to the actual; he was, in a style we can recognize as modern, both bookish and autobiographical." Here was a young author who would fill the opening pages of *Moby-Dick* with scores of obscure references to whales, and now he had found a young woman fully capable of playing literary games right along with him in perfect sympathy. For more conventional lovers, such games would be dull, but for this couple they were essential to their relationship—a kind of elaborate foreplay that would be all the more exciting because no one else would understand what they were doing.

The idea that they dress themselves as characters from *Humphry Clinker* probably originated with Herman, whose surname appears in the book. On his travels, Tabitha's brother has an amusing encounter with the good-hearted Count and Countess de Melville and their stunningly beautiful goddaughter, Seraphina Melvilia. How could Herman resist this novel, with its echo of his name and Sarah's in Smollett's invention of Seraphina Melvilia? In his own work Melville had already raised the possibility that Smollett's tales were excellent for inspiring amorous thoughts. In his second book, *Omoo*, he suggests that reading Smollett may have so aroused one of his male characters that the fellow spent the next seven days trying to seduce a reluctant South Seas maiden.[3]

In any case, once a big man like Melville clothed himself in the exotic finery of the old Middle East, he would make a strong sexual statement for anyone in his culture. By the prejudices of the time, he was an intimidating figure in Turkish garb, suggesting the forbidden ways of the harem and the supposed ferocity of the old Ottoman warriors. If that weren't enough to keep Sarah amused, he went a step further. He decided to play a risky prank on one of Evert Duyckinck's friends who was passing through town. Sarah was expecting that one of the guests that night would be arrayed as a bride. (In *Humphry Clinker* a funny episode tells of a "debauched" servant who "robbed" a man of his not-so-innocent bride.) So, when Melville heard that Evert's friend had just been married in March, and that the wife was accompanying him on their present journey, he saw a chance to do something no one would forget. He would play two parts inspired by Smollett's novel—the Turk and the "debauched" abductor—and he would begin by stealing a real bride and bringing her as a kind of trophy to the party.

ON AUGUST 9, at three in the afternoon, twenty-five-year-old William Allen Butler and his bride, Mary, arrived at the Pittsfield depot, where they were supposed to change trains after a short delay. Melville was waiting for them and hopped aboard just as "the last echoes of the engine whistle" announced their arrival. They had no idea of what was to come, but they weren't the sort of pair to indulge a prankster. Butler was a serious young man with a long, solemn face framed by bushy sideburns. His wife was a small woman with prim features and a severe hairstyle parted in the middle and tightly bunched on either side.

Though Melville barely knew Butler—a minor contributor to Duyckinck's highly regarded *Literary World*—he walked straight up to the young man's wife and, without a whisper of explanation, led her off the train and into a waiting buggy. In the confusion of their arrival, and with their friend Duyckinck in view, the couple must have assumed that the tall stranger was there to help. Before Butler could say a word, Melville cracked the whip and rode away with the startled woman at his side. As the incensed husband later told the story, he was so taken off guard by this that he simply watched in disbelief as his wife was "whisked out of the cars by a strange man with luxuriant beard . . . and whirled away in a buggy behind a black pony of very questionable build, gait, and behavior."

Then, in a panic, Butler gave chase, and with Duyckinck at his side, he was able to track down the bearded desperado to the supposed hideout—Broadhall—where the "abducted" bride was being held. The idea was that Melville wanted to lure the couple into abandoning their trip and agreeing to stay for the party. He hoped they would play along, so that he could surprise Sarah by showing up with a pair of prisoners from a passing train to dazzle one and all. Or, as

Butler later put it, "The object in view was to capture us for a masquerade which was on the carpet for the evening."[4]

The seething husband was truly alarmed by the prank and wanted nothing to do with Melville. Refusing to stay at Broadhall, he stormed off with his wife, and was still fuming about the incident months later, when he would get his revenge in a Washington newspaper by taking a few literary shots at the author in a review of *Moby-Dick*. He could have made even more trouble if he had wished—he was a lawyer and knew his rights in court, and he had a powerful father in New York, Benjamin Butler, a former attorney general of the United States. "We must enter our decided protest," the angry husband would say of Melville in his review, "against the querulous and cavilling innuendoes which he so much loves to discharge, like barbed and poisonous arrows, against objects that should be shielded from his irreverent wit." Though these words referred to Melville's writing, Butler must have been thinking that his wife was one of those who should have been "shielded" from the author's "irreverent wit."[5]

This episode soon became the talk of the party that August evening, especially because Butler went away complaining that Melville, the bearded Turk, had behaved like a Barbary pirate. "We are not to be caught in this Berber fashion," he declared. Such an irresponsible bit of drama on Melville's part must have left his New York friends shaking their heads in disbelief. They had never seen anything like it from him. Evert Duyckinck had championed his career from the beginning, and the two men had grown close over the past few years. Only three years older than Melville, Evert shared many of his literary interests, and was impressed by the depth of his knowledge and the discipline of his writing habits. He was drawn to the light side of Melville's early work—what he called the author's "fancy-sprinkled page." But this unsettling prank seemed to come out of nowhere. Never again in his long life would Melville do anything in public as wild and bold as his "kidnapping" of a bride—unless we see *Moby-Dick* itself as the next shocking performance that would arise from what Butler feared was a touch of the Barbary shore in the Berkshires. Under Sarah's influence, the man who had married a judge's daughter would find new inspiration for breaking rules in his work and offending respectable people.[6]

EARNEST, DILIGENT, AND RESPECTABLE, Duyckinck was closer in temperament to William Allen Butler than to Melville. His influence as a critic and editor was so high in New York that cultivating his friendship made sense for any young writer, and so Melville was inclined to overlook their differences. If Evert thought that his friend had lost all the high spirits of those earlier days at sea, this episode with Butler's bride must have shown him otherwise. Walt Whitman—an expert on high spirits—was one writer who never sought Evert's approval, and who always thought he was an empty suit. He considered him priggish and pretentious, and the same went for the editor's younger brother and colleague at the *Literary World*, George Duyckinck. "I met these brothers," recalled Whitman in old age. "They were both 'gentlemanly men'—and by the way I don't know any description that it would have pleased them better to hear: both very clerical looking—thin—wanting in body: men of truly proper style, God help 'em!"[7]

As Melville would later discover to his sorrow, Evert could turn on a friend at a moment's notice, but for now he was having too much fun to ponder the nature of their relationship. In deference to Evert,

Butler probably decided not to make a greater fuss. Yet what a risk Melville had taken, and all for nothing except a few laughs. What if Melville had staged the prank all along simply to give Sarah a pointed message about husbands and wives? She may have been eager to cross the line into adultery, but was he?

They never lacked for secluded spots where they could meet on the 250 acres of the farm. As Evert would observe during his stay at Broadhall, "Herman Melville knows every stone & tree." If they soon found themselves alone in a quiet corner of the mansion or in the barn or the woods, would Melville seize the moment and give himself to her? His "kidnapping" of poor Mary Butler was perhaps the answer. Taking a woman away from her husband, whatever the reason, was no joke in his world—unless it was his bold way of proving to Sarah and himself that he could cross that line if he wanted to.[8]

BY THE TIME THE COSTUME PARTY BEGAN, Butler and his wife were gone and on their way elsewhere, and the evening was allowed to unfold without any more drama. The event was a great success. It was a proper ball with dancing and a midnight supper. The last guests didn't retire until after one in the morning, and when Sarah went to bed, she was pleased with her efforts. More important, she didn't seem upset at the Turk who dutifully showed up to entertain everyone with his fierce looks and perhaps a suggestive swagger—though without any prisoners in the end. (The party had to make do with a costumed bride.)

It may not have been apparent to all the guests, but Aunt Tabitha and the Turk were quite a potent match. No worse for the late night, they were on the road again the next day with some of their friends and family, dragging them off to an elaborate picnic deep in the countryside. Unused to so much merriment, Duyckinck complained that "the inexhaustible Mrs Morewood . . . understands the art of making a toil of pleasure."

Herman drove a big wagon full of picnickers and provisions while Sarah led the way on a colt named Black Quake, a rambunctious horse with a name that suggested the thudding force of its galloping speed. It was the kind of mount Sarah preferred, a little dangerous but full of life, and she was thoroughly capable of handling the creature. With a few others on horseback at her side, she rode down a narrow country lane called the Gulf Road, which led into a steep wooded area north of town. The group stirred up so much dust at a gallop that they looked like "a flight of Cossacks in the desert." Lizzie Melville was along on this excursion, but perhaps because she was feeling envious and overlooked, she made the mistake at one point of mounting a horse.[9]

Unable to match the skills of her rival, she took a tumble when the horse threw her. Herman came to the rescue, but she was unharmed, and she tried her best to carry on as if nothing had happened. It was an awkward moment, and not a good sign for Lizzie, who had been a new bride herself only three Augusts ago. If he had not considered it before, Melville must have started wondering now whether he had married too soon. At any rate, the picnic went ahead, the woods were explored, and by the time they all returned to Broadhall it was almost dark.

6

THE DOCTOR'S REPORT

Sarah Morewood's seductive powers worked their wonders on more than a few men, including one who, in the 1850s and many decades beyond, was more famous than Herman Melville. From 1848 to 1855 Dr. Oliver Wendell Holmes spent his summers in Pittsfield at a pleasant retreat near Broadhall, in the meadows along the Housatonic. Usually accompanied by his wife and children, he was then in his early forties, a celebrated poet, Boston wit, and renowned dean of the Harvard Medical School. Though not an easy man to impress, he was always fascinated by Sarah, and a little unnerved by her provocative glances and easy grace.

Half in love with her, but too cautious to make love to her, Dr. Holmes observed Sarah closely in his years as her neighbor, and he left behind elaborate descriptions of her that are almost clinical in their detail. He was quick to notice a sensual air to her movements, a delicate quality that gave him the impression that she glided, rather than walked. "The clover hardly bent beneath her," he said in a poem written for her. Likewise, in *Elsie Venner*—his novel inspired by Sarah and published a decade later, in 1861—he described the book's heroine as "slender . . . with a peculiar undulation of movement, such as one sometimes sees in perfectly untutored country girls, whom Nature, the queen of graces, has taken in hand."[1]

This dark, exotic "wild beauty" leaves a cloud of dust as she gallops over Rockland (Pittsfield), where she is often just a fleeting vision in the shadows of its hills and its largest mountain, a towering mass like the real Mount Greylock. "Elsie loved riding," Holmes writes, and her "wandering habits" leave everyone in the neighborhood of her father's eighteenth-century mansion questioning what she does in "her sylvan haunts." Her "stately old home" in the countryside, with its "broad staircase," its two large parlors, its "Old-World notions of strength and durability," and its wide view of the distant "blue mountain-summit," is unmistakably Broadhall. In fact, early-twentieth-century guidebooks would sometimes point out that the novel was largely set in "what is now the Pittsfield Country Club."[2]

Though Holmes added a lot of unnecessary speculation about a mystical source of Elsie's strange charms, the story captures much of Sarah's life in the Berkshires. Elsie is a strong-willed young woman of means who shocks her idyllic community with her desperate search for love. One of her admirers calls her "a wild flower" with "marks of genius—poetic or dramatic—I hardly know which." The most obvious source of her appeal is her magnetic gaze—"black, piercing eyes" that can stop a man with a single glance. Her appearance has a startling effect on almost everyone: "She was a splendid scowling beauty, black-browed, with a flash of white teeth which was always like a surprise when her lips parted." Holmes found her looks so haunting that he included in a late chapter the blunt declaration, "Elsie would have been burned for a witch in old times."[3]

Dr. Holmes had roots in Pittsfield that ran even deeper than Melville's. As he was proud of saying, "All of the present town of Pittsfield, except one thousand acres, was the property of my great-grandfather." If anyone doubted it, he kept Jacob Wendell's deed in the entry to his summer place. It was dated 1738. His present house stood on the last remaining acres, a mere 286 of the original 3,600. After she was settled in Broadhall, Sarah would occasionally visit him, and he relished his time with her. His own wife, Amelia, was, like Melville's, the daughter of a Massachusetts Supreme Judicial Court justice, but she was so painfully shy and retiring that few of the neighbors knew much about her.[4]

As Sarah left his house after a visit one day, Holmes had to steady his nerves when she paused before her horse and asked him to lift her into the saddle. He leaned over to take her boot in his hand, her skirts rose, and the great physician began to tremble. She had that kind of effect on men, and she knew it. "Now, gallant, now! Be strong and calm," he recalled telling himself, "her foot is in thy hollowed palm." In those days just the sight of a woman's shapely ankle was enough to excite most men. To hold it in your hand—as Holmes's language indicates—was pure delight. The moment was so titillating that he imagined even the surrounding flowers were shocked: "The violets shut their purple eyes, / The naked daisies stared in wonder."

After watching Sarah bid him farewell and race off in a burst of speed—with her favorite cloth cape no doubt flying in the wind—he decided to call her Camilla. In his dreamy vision, she resembled the swift beauty in the *Aeneid,* who outstripped "the winds in speed upon the plain," and who waged war fearlessly with one breast exposed in battle. When Camilla is on horseback, every eye is on her. As John Dryden wrote in his translation,

> Men, boys, and women, stupid with surprise,
> Where'er she passes fix their wond'ring eyes:
> Longing they look, and gaping at the sight
> Devour her o'er and o'er with vast delight.

Sarah couldn't have doubted that she struck a nerve in the dean of the Harvard Medical School. Within days of that visit, he sent her a poem full of extravagant praise for her beauty and charm—especially recalling her "sparkling" smile and the lingering thrill of "her foot" in his hand. Confident that she would understand the reference, he gave the poem a title of only one word—"Camilla."[5]

AS ONE OF THE MANY who fell under Sarah's spell, Dr. Holmes was both amused and alarmed by the close relationship that developed so quickly in 1850 between his new neighbors, Mrs. Morewood and Mr. Melville. He came to know his fellow author well in the next few years, and at one point he "attended & prescribed for him" in a professional capacity. Moreover, he knew Judge Shaw very well. Because he had been an expert medical witness in the judge's courtroom, he fully appreciated that Shaw was no man to be trifled with. It didn't take much to imagine how furious the chief justice would be if his son-in-law became ensnared in an adulterous affair.[6]

Adultery was, in fact, a hot topic in 1850. Hawthorne's *The Scar-*

let Letter, which came out in March, was selling briskly, and the New York press was full of lurid stories that summer about an affair between the writer Nathaniel Parker Willis (whom Melville knew) and the wife of the actor Edwin Forrest. One night in June the actor attacked Willis in Washington Square, throwing him to the ground and "thrashing" him with a whip for "interfering in his domestic affairs." Forrest's warning to onlookers, after it was published in the newspapers, must have sent a chill down the spine of many New Yorkers contemplating an affair with someone else's spouse. "Gentlemen, this is the seducer of my wife, do not interfere!" he shouted, while Willis lay on the ground "screaming . . . 'Help! Help! Save me! Police! Take him off!' " Willis managed to escape, but the resulting scandal ruined both men's careers.[7]

The intense love affair between Herman and Sarah that took root that same summer was never much of a secret to Dr. Holmes, who had the confidence of many in the neighborhood, and who closely followed events in the small town. It intrigued him to see how the relationship would unfold, and whether the dangerous game being played by the "wild beauty" and the American Robinson Crusoe would turn tragic or comic. Holmes was also a little jealous of Melville, who was not only a much better writer, but also more physically imposing. As the doctor once admitted, he himself was so short that he measured "exactly five feet three inches when standing in a pair of substantial boots." In most situations, he would always have been looking up at Melville, who was almost five feet ten inches. "A dare-devil fellow," he would say of his Melville character in *Elsie Venner* when he fictionalized some of the events of the early 1850s. As a pun on the whaling epic that his neighbor wrote in Pittsfield, he would name Elsie's closest companion Dick. In the novel the young man is "handsome and romantic," but too reckless, and too much like Elsie.

After surviving a youthful period of dangerous travels overseas among "savage" primitives, Dick comes home to live in the ancestral mansion to which he and Elsie both have claims. "A young man of remarkable experience for his years," he captures the interest of the little town as an "adventurous" figure "with stories to tell." Elsie's beauty tempts him like "forbidden fruit," and her father tolerates his presence because "anything that seemed [likely] to amuse or please Elsie was agreeable to him." Melville's prank of abducting a bride and dressing as a Turk seems to have been known to Holmes, who describes Dick as a lusty young man with eyes like those of a Turkish pasha "in the habit of ordering his wives by the dozen." Soon the two young people are together everywhere in Rockland. "Somehow or other," writes Holmes, "this girl had taken strange hold of his imagination." A neighbor in the novel asks, "Have you seen them galloping about together? He looks like my idea of a Spanish bandit on that wild horse of his."[8]

Holmes drew this image of Melville, like many other things in *Elsie Venner*, from local lore. It seems to have been one of Melville's eccentricities while he was in the Berkshires to go out riding and greet others in Spanish. Nathaniel Hawthorne, then temporarily living six miles away in Lenox, was given the full treatment one day when he was outside reading the newspaper at his cottage and Melville rode by. "A cavalier on horseback came along the road, and saluted me in Spanish; to which I replied by touching my hat, and went on with the newspaper. But the cavalier renewing his salutation, I regarded him more attentively, and saw that it was Herman Melville!" Clearly, something—or more accurately someone—"had taken strange hold" of Herman's imagination.[9]

WHEN *ELSIE VENNER* APPEARED a decade after *Moby-Dick*, Sarah told an old friend that it "created a Storm in many quarters" of Pittsfield, but that Holmes—who had moved away by then—didn't care about "the strife of winds" he had stirred up. The storm soon died out, and the novel has long since been forgotten. The book, however, does offer a revealing glimpse into Melville's secret life by a physician who saw it as something of a case study worth documenting. He may not have been an accomplished novelist, but as a scientific observer he had one of the sharpest minds of his generation, contributing to advances in germ theory, antisepsis, and anesthesia (a term he coined).[10]

Of all the intimate stories Holmes had uncovered in his long medical career, this one fired his imagination more than any other, and over the years he simply couldn't resist the temptation to smuggle some of the truth to the larger world beyond the Berkshires in the form of *Elsie Venner*. The novel isn't really a love story, but rather a tale of characters searching for love and willing to do anything for it. As Holmes sees it, Elsie and Dick are doomed from the start by their refusal to live by the rules of their community, and the good doctor is at pains to portray them as victims of their own disease of discontent. At the risk of getting a little ahead of our tale, it is worth noting that when Dr. Holmes left Pittsfield in 1855, he seems to have given Melville a friendly warning, advising him at that late point in the affair that it wasn't such a well-kept secret and might cause harm. In turn, Melville himself did a little smuggling of this truth into a story published not long after Holmes attended him that year, when—according to a mutual friend—the doctor treated him "with fraternal tenderness . . . giving him his best medical advice."[11]

In his story "I and My Chimney," written long after Melville had decided to make his home next to Sarah's, he transformed Holmes's medical examination of him into a darkly funny tale about an architectural expert called in to examine an old farmhouse chimney for "soundness." The wife of the house wants to get rid of the ancient chimney, but the husband is fond of it despite its age and structural faults and forbids anyone to disturb it. The expert—a Mr. Scribe—descends deep into the cellar to study "the root of the matter," and concludes that something is amiss. "It is my solemn duty to warn you, sir," Mr. Scribe reports, "that there is architectural cause to conjecture that somewhere concealed in your chimney is a reserved space, hermetically closed, in short, a secret chamber. . . . How long it has been there, it is for me impossible to say. What it contains is hid, with itself, in darkness." The expert refuses to say what should be done with the chimney, but he strongly suggests that the secret chamber has made it unsound, and that—whatever action is taken—his own conscience is eased by telling the husband the truth. He concludes his report with the words, "Trusting that you may be guided aright, in determining whether it is Christian-like knowingly to reside in a house, hidden in which is a secret closet."

The husband is unmoved and refuses to change a thing. He pays Mr. Scribe fifty dollars to alter his diagnosis and issue a certificate saying that "you, a competent surveyor, have surveyed my chimney, and found no reason to believe any unsoundness; in short, any—any secret closet in it." Proudly, the husband frames the certificate and attaches it to the fireplace in the hope that his wife will stop questioning the future of the obviously phallic object. "I and my chimney will never surrender," he vows, but the dispute is still raging when the story ends.

Holmes may never have seen Melville's thinly veiled criticism of his honest evaluation. It appeared anonymously in *Putnam's Monthly* in 1856 and wasn't published under Melville's name until after both men were dead. Unlike Mr. Scribe in the story, Holmes waited a few years and issued a more detailed report in the disguised form of his novel. He called *Elsie Venner* "a psychological romance," but even a doctor as distinguished and advanced as Holmes couldn't fathom the depths of the "strange hold" that Sarah so quickly established over Herman. It isn't surprising that when the doctor's son—the great Supreme Court justice Oliver Wendell Holmes Jr.—recalled in 1921 his childhood memories of Melville, he described him as a mysterious "rather gruff taciturn man" who wrote "with an actuality and first hand contact with life that I suppose hardly pleased my father's generation as much as it does me."[12]

The "chimney" was never entirely metaphorical. It still dominates the farmhouse that Melville acquired at the end of the summer of 1850. In the cramped and dimly lit cellar where the "root" of the chimney sits, the walls are now lined with boxes of manuscripts belonging to the Berkshire County Historical Society. Some of the research for this book was done in that cellar, especially with the contents of a box containing, ironically enough, one of the largest collections of letters by Sarah Morewood.

7

THE SCARLET ESSAY

There was another friend besides Mrs. Morewood who brightened Melville's first summer in the Berkshires. Only a couple of months before Herman began his holiday in Pittsfield in 1850, Nathaniel Hawthorne moved with his family to the small farmhouse he rented near Lenox. A replica of his little red cottage stands in its place now, but the general view toward Stockbridge is more or less the same. That summer, at forty-six, Hawthorne was enjoying a surge in popularity with the success of *The Scarlet Letter,* the first edition of which had sold out in only two weeks when it appeared in March. Among the novel's early admirers was Sarah Morewood, who—naturally— was fascinated by the subject of the book, and in particular by the bright image of the *A* on Hester Prynne's breast. She was so drawn to the story that her sharp eye spotted one day a newspaper reference to the German-American artist Emanuel Leutze planning to paint a scene from the novel. Leutze, best known for his *Washington*

Crossing the Delaware, claimed that he had seen an old painting at a German castle depicting a woman remarkably like Hester wearing a scarlet letter, and that his work would be essentially a copy of this earlier one.

Sarah was intrigued to think that centuries ago, an artist had created an image of a woman in the same situation as Hester. It gave more validity to Hawthorne's vision of the brave mother of little Pearl, and she was especially fascinated by the claim that the original work was found in a place called the Castle of Pearls. But, as she told a friend in New York, *The Scarlet Letter* was so powerfully conceived that a whole set of paintings could be based on its scenes. "I can imagine," she said, "that almost any artist might paint well from Hawthorne's descriptions—they are so vividly drawn."[1]

When Melville and Hawthorne were not as highly regarded as they are now, Sarah Morewood saw the greatness in both men, and could share with Herman the enthusiasm that overwhelmed him when he became friends with his fellow writer. Later generations have marveled at the fact that two great American authors at the height of their powers would forge a brief friendship when they lived near each other for a year in the Berkshires. What has been absent from this story is the part played in it by Melville's love for Sarah.

AT THE OUTSET, Melville was quick to appreciate that he and the novelist on the other side of Lenox had a lot in common. By different means, and in much different styles, they were aiming for a similar artistic goal. In *The Scarlet Letter* Hawthorne identified that goal as the "neutral territory . . . where the Actual and the Imaginary may meet, and each imbue itself with the nature of the other." In this way Hester's tale can "imbue" the actual evils of the old Massachusetts Bay Colony with the real horror and pathos that time has worn away. For Hawthorne, fiction enabled the letter of shame to burn once again, which is why he pretends at the beginning of the novel that he discovered an actual scarlet *A* in the dusty attic of Salem's Custom House, and that when he placed it to his breast, he felt a "burning heat . . . as if the letter were not of red cloth, but red-hot iron." For both Hawthorne and Melville the past wasn't dead as long as the imagination could rekindle it.

Yet it was partly bitterness and spite about his own past that drove Hawthorne into making the Berkshires a temporary refuge. A change in political administrations in Washington had cost him his job at the Custom House in his native Salem. He placed much of the blame for this setback on his political opponents locally, and—believing he was unappreciated—he vowed "to remove into the country and bid farewell forever to this abominable city." Perhaps identifying too closely with Hester Prynne's tragedy in his masterpiece, he saw himself as someone unjustly shamed by his community, and he turned angrily against it—as if he shared Hester's punishment of having to wear a mark of disgrace in public. "I detest this town so much," he wrote only a month before *The Scarlet Letter* was published, "that I hate to go into the streets, or to have the people see me." By moving to Lenox at the opposite end of Massachusetts, he went about as far as he could go without actually leaving the commonwealth. His new place was a good walk beyond the town itself, and in his relative isolation, he could avoid seeing other writers. It didn't take long for his path to cross Melville's, however, partly because the younger writer was emerging from his own bubble of isolation to enjoy the reinvigorated social scene at Broadhall.[2]

THE OCCASION OF THEIR FIRST MEETING was a rural outing on August 5, 1850, a day that would become a famous set piece in literary history. In a group that included Dr. Holmes and Evert Duyckinck, along with a few other visitors from New York and Boston, Hawthorne and Melville came together for a summer hike to the top of Monument Mountain, a few miles south of Stockbridge.

As a climb, it wasn't too demanding and gave ample opportunities for the group to converse, joke, and horseplay as they went along. The highlight came when a sudden thunderstorm forced everyone to take cover, and a mug of champagne was passed around to raise the group's spirits as they sheltered under a cliff. Later, at the summit, Melville couldn't resist a new opportunity to show off, displaying his sailor gymnastics, straddling "a peaked rock, which ran out like a bowsprit," and pretending to haul up an imaginary rope. (Wary of the rocky ledges, Dr. Holmes "peeped about the cliffs and protested it affected him like ipecac.")[3]

The group remained together after they descended the mountain, and then they all shared a long dinner. This eventful day gave Melville and Hawthorne several opportunities to size each other up, and each was intrigued by the other. In the days that followed, the younger writer received an invitation to pay a visit to the older one. It was rare for Hawthorne to warm so quickly to a new acquaintance, but his self-imposed exile from Salem had lasted long enough to make him eager for a little companionship from an author with a background as interesting as Melville's.

Hawthorne wasn't the only one drawn to this new friend. Hawthorne's wife, Sophia, found Melville charming right from the start. She could barely contain her enthusiasm after his first visit to the red cottage. A handsome, creative woman ten years his senior, she thought Melville was "a man with a true warm heart & a soul & an intellect—with life to his fingertips." She also found his physical presence captivating. Her husband's good looks had a delicate, refined quality, but the former sailor with his full beard and strong body had "an air free, brave & manly." Most of all, she was moved by the power of his eyes. His gaze "does not seem to penetrate through you," she observed, "but to take you into himself."[4]

Fascinated by Hawthorne's books, his character, and his summer life in the Berkshires, Melville decided on an impulse to write an essay about him. He created it with amazing speed one weekend at Broadhall, and in the second half of August, Evert Duyckinck published the essay in the *Literary World*. At first the identity of the author was a mystery. The journal identified him by the misleading but mysterious title "A Virginian Spending July in Vermont." The writing was so extraordinary, and the praise so extravagant, that people were convinced it was the work of a major critic temporarily hiding behind a pseudonym. It was obvious that this "Virginian," who was supposedly idling the summer away in New England, was a superb stylist whose comments soared to lyrical heights few literary journalists could touch. (And this "Virginian" seemed to have little respect for ordinary reviewers, joking, "There are hardly five critics in America; and several of them are asleep.") Pleading for American readers to acknowledge the genius of Hawthorne, Melville—the mystery author—argued that the essence of the nation was embodied in the works of the man: "The smell of your beeches and hemlocks is upon him; your own broad prairies are in his soul; and if you travel away inland into his deep and noble nature, you will hear the far roar of his Niagara." The portrait that emerges from this essay is of an American Shakespeare, a literary giant achieving his grand effects not by imitating Shakespeare but by exploring some of the same tragic aspects of human experience. "I do not say that Nathaniel of Salem is a greater than William of Avon," wrote Melville, "or as great. But the difference between the two men is by no means immeasurable. Not a very great deal more, and Nathaniel were verily William."[5]

Generous as it was, the enthusiasm here was not entirely for Hawthorne's benefit. At this stage, Melville didn't know the author's work that well, but he had recognized a kindred soul—both in the pages of the books and in the conversation of the man—and he wanted to use the achievement of the older writer to justify the high ambitions of his own work.

The essay is a plea to America to take its best writers seriously, and to give them the respect they deserve: "Let America then prize and cherish her writers, yea, let her glorify them. They are not so many in number, as to exhaust her good-will." Sounding like a literary evangelist, Melville declares, "Believe me, my friends, that Shakespeares are this day being born on the banks of the Ohio." The country was preparing for the day when it would enjoy "political supremacy among the nations," but it was "deplorably unprepared" for the coming time when its writers would rank among the world's best. One day, Melville predicted, it will not be necessary to use Shakespeare or any other British writer as the measure of an American writer's greatness. "We want no American Goldsmiths, nay, we want no American Miltons. . . . Call him an American, and have done, for you can not say a nobler thing of him."

Ahab and Ishmael, like Hester, would be the faces of an American literature second to none. In his own study of British writers and artists, Melville had learned how to reach for a higher form of art, but the literary and historical models were only starting points. It would be up to this young man barely past thirty to find the unique shape and texture of *Moby-Dick*. His ambition at this moment knew no bounds. "If Shakespeare has not been equaled," he boasted, "he is sure to be surpassed, and surpassed by an American born now or yet to be born." In all his dreams of immortal fame Shakespeare himself could not have allowed his spirits to soar any higher than Melville's in this inspired essay written in the romantic landscape of a summer in the Berkshires.

Yet there was something in the darker side of Hawthorne's writing that also held a powerful attraction for Melville. He had lately been reading the author's collection of stories *Mosses from an Old Manse,* and had been drawn to "that great power of darkness in him [which] derives its force from its appeals to that Calvinistic sense of Innate Depravity and Original Sin." His copy of the stories has survived, and in Hawthorne's "Young Goodman Brown," Melville underlined this telling remark: "It shall be yours to penetrate, in every bosom, the deep mystery of sin."[6]

Sarah Morewood and Herman Melville didn't need Hawthorne's books to guide them in penetrating "the deep mystery of sin," but that aspect of his work was—above all else—the most alluring part of his appeal for them. The exact moment when Melville wrote his essay on Hawthorne is important, for it was in that hectic August weekend of Sarah's costume party, when he was "kidnapping" the wholesome bride Mary Butler, that he dashed off the essay. It is a long and, at times, almost breathlessly exuberant piece written with such speed and concentration that the intensity of Melville's excitement is palpable on the page. Reading Hawthorne may account for some of that excitement, but the greater part must have come from the same attachment that made Melville dress like a Turk, steal a wife from a train depot, and stay up until one in the morning with a young Aunt Tabitha eager for romance. What he found in Hawthorne's work gripped his imagination because it mirrored what he was suddenly finding in his own life—a sense of passion and adventure with a strong undercurrent of something forbidden and secret.

Though Melville wouldn't resume an intensive routine of writing his new novel until the autumn, it is worth noting here what Ishmael says in *Moby-Dick*: "I love to sail forbidden seas, and land on barbarous coasts." So did Melville, and it must have thrilled him to the core to discover that he had sailed all over the world only to discover that his beloved Berkshires contained primitive delights as wondrous as those in the Pacific. As he was falling in love with Sarah, and contemplating a voyage into an unknown world forbidden by every authority in his society, it is little wonder that he found so much to like in the author of *The Scarlet Letter*. He put it plainly to the unsuspecting readers of the *Literary World*. "It is that blackness in Hawthorne, of which I have spoken, that so fixes me and fascinates me."

Equally unsuspecting were Hawthorne and his wife when they read the essay just after its publication. Carried away with all the praise for her husband, Sophia was practically shaking with delight. Her more subdued husband tried to take all this adulation with some show of modesty, saying that it was "more than I deserve," but Sophia couldn't "speak or think of any thing [else]." As she read the piece over and over, she was increasingly amazed that any critic had dared to shout Hawthorne's praises to the sky with no apologies or reservations. Not yet having any idea of Melville's authorship, she was desperate to discover who could have written such a brilliant essay: "Who can he be, so fearless, so rich in heart, of such fine intuition? Is his name altogether hidden?"[7]

THE HAWTHORNES INVITED Melville to stay at their cottage for a few days in early September, and during that visit either they were able to tease the secret out of him or Duyckinck revealed his identity in a letter to them. When Sophia found out the truth, she seemed genuinely surprised. "We have discovered who wrote the Review in the *Literary World*," she informed her sister. "It was no other than Herman Melville himself!"

Once he was found out, Melville pretended that he had written the whole thing at a time when he never dreamed that he would actually meet Hawthorne. Sophia accepted this white lie wholeheartedly, and gushed to any and all that their new friend was the best possible company. She had long conversations with him, and recalled one evening in particular when there was "a golden light" over the landscape as they sat outside. She was enchanted by his high spirits and found him "a person of great ardor & simplicity. He is all on fire with the subject that interests him. It rings through his frame like a cathedral bell." That phrase "on fire" was not one that most people who knew him in earlier days would have applied to Melville. He was usually more restrained, but not now. The sound of that bell was from something less spiritual than a cathedral.[8]

Hawthorne was more cautious in his response to this fervent admirer. He was used to working alone, and his darker, brooding side did not align with Melville's almost boyish enthusiasm for a new age of American literary excellence. He may have also suspected that his admirer's praise was, for whatever reason, overdone, so he hesitated to share more of his life with the younger man, and was measured in his responses to any questions. Often he said nothing at all, but just nodded or exchanged meaningful glances as Melville spoke to him in long, rolling tides of conversation.

Unabashed, the younger writer found excuses to see this "silence" as something positive, telling Sophia that her husband's "great but hospitable silence drew him out—that it was astonishing how *sociable* his silence was." With witty conciseness, Dr. Holmes summed up Hawthorne's character in one memorable couplet: "Virile in strength, yet bashful as a girl, / Prouder than Hester, sensitive as Pearl."[9]

8

"HOLY INFLUENCES"

To repay his hospitality at Lenox, Melville invited Hawthorne to dinner at Broadhall on a Wednesday in the first week of September 1850. Though Hawthorne was the more esteemed, Melville had the old mansion at his disposal and could entertain the older writer much more grandly. The sale of the house to the Morewoods would soon be finalized, but Melville was acting as though the place belonged to him. That had been clear in August to Evert Duyckinck, who noted, "Melville . . . treated the house as his own & would suffer no payments." For a short while at least, both Sarah and Herman acted as if they owned a mansion that still formally belonged to neither of them.

Time was running out. Rowland Morewood was planning to leave for England on October 9 to visit his family, and he wanted to complete the sale so that renovations could start while he and Sarah were away. His wife, however, didn't want to go with him. Only "reluctantly"—as she put it later—did she finally give in and agree to the voyage. Later, in a sarcastic understatement, she would recall her mood after her bitter surrender: "I did not feel the most happy person in the world." She didn't have much choice. In England, Rowland's father, whose wealth helped to sustain the New York branch of the family business, was eighty-six, and was unlikely to live much longer. The family expected Rowland and his wife and child to pay a visit before the patriarch was gone. It wasn't acceptable that he would come by himself, so Sarah must have known all along that she couldn't back out. Yet the more she stayed in the Berkshires, the harder it was to leave Broadhall and Melville behind. The only consolation was that when spring returned next year, she would be back, and Broadhall would be hers. She was already making plans to become a bright and permanent fixture on the social scene of the area. While Melville was getting to know Hawthorne, she applied her own modest literary talent to writing a poem for the biggest civic event of the year in Pittsfield.[1]

On September 9 the whole town was going to march through the streets to celebrate the dedication of the new cemetery. Several thousand people were expected to turn out, and all the leading citizens would be gathered in a central grove for speeches, prayers, and songs. Dr. Holmes had agreed to read a long poem, and Sarah wanted to submit her verses to the choir in the hope they would be set to music. At a time when women writers often struggled to get their work into print, this kind of civic occasion gave Sarah an opportunity to receive some recognition for her talent, so it came as a delightful surprise when her submission was accepted along with that of another woman. She was identified in the program as "Mrs. J.R. Morewood of New York, a Lady who is about to become a resident of Pittsfield." The other woman—in keeping with the more accepted standards of female modesty—was described as simply "a Lady."[2]

It must have been one of the great moments in Sarah's young life when the choir sang her "Ode" before an audience of four thousand on a nearly perfect late summer morning under the blue skies of the Berkshires. In the local view, this triumph established her as one of the town's literary figures. The *Pittsfield Sun* would later describe her as "a lady of superior literary accomplishments." For a woman in a small town, she couldn't have hoped for better praise. It was certainly more than Emily Dickinson ever received in her lifetime in nearby Amherst, Massachusetts.[3]

The strange fact that a seductive woman like Sarah would be credited with writing a hymn didn't escape the notice of Dr. Holmes. "What the diablo had Elsie to do with hymns?" his narrator asks incredulously in *Elsie Venner* when a well-thumbed hymn book is discovered in her room. Unlike the conventionally religious verses of the other "Lady," Sarah's lines were pure pantheism, mentioning neither God nor heaven, but locating all beauty and spiritual power in Nature. Such was her devotion to the natural world that she praised its "holy influences" as the one sure source of grace and comfort. Nature was immortality, and whether as a woman or leaf or stream, all things were united in the eternal cycle of the seasons, and of life and death. Every path was circular, every living thing a marvel inhabiting a world of ceaseless change.[4]

> The stream whose waters glide along,
>   Till lost amid the rolling sea,
> Shall tell us of the eager throng
>   Fast hurrying to eternity.
>
> But spring unfolds a sweeter tale,
>   From which the heart may comfort learn,

When flower-gems strown o'er hill and vale
  Proclaim the op'ning year's return.
. . . .
Then "Woodlawn!" hallowed be thy ground,
  We consecrate thee to the dead;
Rest they, where Nature all around
  Her holy influences hath shed.[5]

Though both her talent and her theme followed predictable patterns, her fascination with the natural world, her great curiosity, and her openness to new experiences meant that she was far more in tune with Melville's ambitions and interests than any other woman he knew. He was as reluctant to let her go as she was to leave. Their first summer together was no casual fling. If Melville had been a typical womanizer of the time—like Alexander Gardiner—he would have welcomed Rowland's planned trip overseas as a timely escape from any entanglement. But Melville wanted to be entangled. When Sarah returned in the spring, he planned to be not only in the Berkshires, but as close to her house as possible without actually moving in.

His plan only made sense in light of his fervent attraction to Sarah. Otherwise, to most people, it would have seemed merely foolish and irresponsible. He shared his idea with Lizzie's father the day after the choir had performed Sarah's verses at the cemetery dedication. Judge Shaw happened to be in the area on court business, and his son-in-law couldn't wait to ask him for help. What Melville proposed was that he and the rest of his New York household should abandon the city immediately and buy a place in the Berkshires. The property in question was a humble old farmhouse on 160 acres bordering Broadhall. To buy it, he needed $6,500, but his finances were in such disarray that he couldn't afford to pay any of that amount. Would Judge Shaw loan him the money for it? he asked.

It was an audacious and impulsive request, especially considering that the judge had his legal duties to attend to. Strictly as a land deal, it wasn't a good idea. The asking price was exactly the same amount as Rowland had paid for Broadhall. For $6,500 Rowland had bought a mansion, and a much better farm that was two-thirds larger than the one Herman wanted. His deal had been a bargain, but the author was being asked to pay the full market value for a mediocre property. To afford it, he would have to go deeply into debt. But without a steady income, he would find it almost impossible to pay off his loans.

To explain this reckless deal, Melville's modern biographers have universally agreed that he was indeed desperate—so much so, in fact, that he was willing to do anything to remain near his new friend and "neighbor," Nathaniel Hawthorne. Buying a place six miles away from Lenox, however, wasn't very neighborly. Renting a cottage, as Hawthorne himself had done, would have made more sense for a former sailor who was never going to be much of a farmer on these rolling acres in Pittsfield. In fact, given the small amount he was earning from his books, renting was the only reasonable plan for staying in the Berkshires. Buying the farm was so beyond his means that—as his most scholarly biographer, Hershel Parker, has pointed out—it would cost him "more money than he had earned from all his five books together, in both England and the United States." By contrast, renting was cheap. As Melville was later to admit, it was possible to find a decent house in town for as little as $150 a year.[6]

There was only one explanation for his desire to buy this particular farm, and for his willingness to pay whatever was necessary to get it, no matter how unaffordable. The young author who thought he belonged at Broadhall with Sarah was making sure he had the next-best thing—a farm adjoining hers. The wonder is that Melville was able to talk Judge Shaw into loaning him $3,000 toward the purchase. The rest of the amount he planned to cover by arranging for a mortgage and a deferred payment to the owner—obligations he was ill-prepared to honor. It was a recipe for disaster, but he couldn't help himself.

THE PURCHASE WAS COMPLETED so quickly that he acquired the farm just four days after Judge Shaw arrived in Pittsfield. Significantly, when the town historian—the poet J. E. A. Smith, a mutual friend of Sarah and Herman—recalled the deal forty years later he let slip that Melville's decision was influenced by Rowland's purchase. "In anticipation of the sale of Broadhall," wrote Smith, "Mr. Melville on the 14th of September, 1850, bought of Dr. John Brewster Sr. the farm adjoining the Broadhall estate."

Often a guest at Broadhall, and romantically linked to one of Sarah's sisters, Smith was present at the "Laurel Wreath" Christmas dinner of 1851. In old age—when both Sarah and Herman were gone—he wrote a long newspaper series on the novelist's life. It caught the attention of Lizzie Melville, then the stoic widow who had steadfastly remained in her unhappy marriage for more than forty years. Pleased with Smith's sympathetic and uncontroversial account, she arranged to have the piece reprinted in a booklet. In addition to correcting some factual errors, she also eliminated an entire section that recalled the day when Mrs. Morewood finally decided that her new house would indeed be christened "Broadhall." In a rare lapse of discretion, Smith had revealed that it was Sarah and Herman who had reached the decision over the name, contriving a sly contest to hide their collusion. Smith used coy language to tiptoe around the truth, but his revealing glimpse of Sarah's relationship with Herman was too much for the widow. It seems the most likely reason she deleted an episode that would appear innocent enough to most readers. Here is the passage:

> One evening in a merry party of men and women more or less distinguished, it was proposed to give [Mrs. Morewood's house] a name; each person present having the privilege of putting one in a basket; the first drawn out to be forever fixed upon the venerable historic mansion. Mr. Melville wrote on his slip the word Broadhall, and that came first to the deft hand [of Mrs. Morewood] which was appointed to be the minister of fate. We have a very strong suspicion that the deft hand was guided by a deft brain, and that so happy a drawing was not so entirely a matter of chance as it purported to be.[7]

Lizzie, the privileged daughter of the chief justice whose world revolved around Beacon Hill, probably wasn't thrilled at the idea of trading her home in New York for a farmhouse in Pittsfield, but she must have agreed to the plan in the end because her doting father wouldn't have loaned the money otherwise. Still, it was difficult to hide just how bad the deal was, and how ominous the future would look if the next book failed. The young novelist would have to fight two battles at the same time, one for art and one for ready cash.

THOUGH MELVILLE MAY HAVE KNOWN for some time that he would try to buy the Pittsfield farm, the suddenness of his action stunned many. One of his cousins wrote to Judge Shaw that she hoped Herman's family "will have no cause to regret" the move. "I confess," she added, "that it surprised *me* at first." Evert Duyckinck and his brother, George, were also caught off guard by the news. "Herman Melville has taken us by surprise by buying a farm," wrote George. "It is mostly woodland which he intends to preserve and have a road through, making it more of an ornamental place than a farm."[8]

There was already a good road running right in front of the farmhouse, connecting it to Pittsfield in one direction and Lenox in the other. If he put "a road through" the "ornamental" woods behind his house, it wouldn't take him any closer to Hawthorne, but it would take him much closer to Sarah's door, which was the idea. Or rather, that was the dream, because he didn't have the money to build a proper road. (The best he could manage in the coming year would be to clear a rough trail good enough for fair-weather travel. It went straight to Broadhall.) Nor did he have the money to build something else he talked about—a tower that would give him a view of Sarah's place over the tree line. When he told Sophia Hawthorne that he was going to construct a new house on his property and include a large tower, she couldn't believe it. He insisted that it was no mere fantasy. "He is really going to build a real towered house—an actual tower," she wrote afterward, as if still trying to convince herself. At the top of that tower, he could have a study and write his books, and always with a view of Broadhall in the near distance.[9]

These were the overexcited dreams that surrounded his purchase, and the only hope he had of seeing them realized was to finish that book of his on the Whale. If he could produce a brilliant book on a gigantic scale that would leave the literary world in awe, then he just might realize the ambitions spelled out in his essay on Hawthorne—

and, not incidentally, earn enough money to keep his dreams alive and enjoy more summers in the Berkshires like this magical one of 1850, now drawing to a close. Success in the coming year would allow him at least to continue dividing his life between Mrs. Morewood and the judge's daughter he had wed before he knew of a better match.

What is perhaps the most outrageous part of his plan is that he was willing to undertake this risky gamble using the judge's money. It was a gesture worthy of someone who had once jumped ship in the middle of the Pacific, abandoning a whaling vessel to try his luck as a castaway. That gamble had paid off handsomely in the end, and now he was hoping to use his pen to write a story that would transform his life once again. For a literary man with such outsize dreams, the stakes couldn't have been higher.

IT IS UNCLEAR when Melville bid farewell to Sarah that year, but both seem to have been in and out of Pittsfield in late September 1850, Melville to organize the move from New York to the new house, and Sarah to prepare for her voyage. By the time Sarah's steamship, the *Niagara*, left for Liverpool on October 9, Melville had already packed up his family's belongings, sent them to Pittsfield, and taken up residence in the farmhouse. It was going to be crowded. His mother, his wife and child, and three of his four grown sisters were all going to live there. With so many dependents continuing to look to him for support, and with his fresh load of debt, a financial success with his new book couldn't come too soon.

Getting the old place in shape would take a couple of weeks. In a long October period of sunny weather, he worked incessantly to prepare the house for winter. After finding some ancient evidence of

Native American culture in the soil of his new property, he decided to call the house and its farm Arrowhead. Almost breathless with exertion and excitement, he wrote to Evert Duyckinck on October 6, "I have been as busy as man could be."

On one glorious autumn day full of red and gold, Melville looked around his new farm and was amazed by the beauty of the landscape. Though Sarah was no longer there, he felt some of those "holy influences" of nature that she cherished. "It has been a most glowing & Byzantine day," he wrote in the evening, "the heavens reflecting the hues of the October apples in the orchard—nay, the heavens themselves looking so ripe & ruddy that it must be harvest-home with the angels."[10]

PART II

—

GREYLOCK'S MAJESTY

With due obeisance & three times kissing of
your Ladyship's hands, & salutes to all your
Ladyship's household, I am,
Dear Lady of Southmount,
Your Ladyship's
Knight of the Hill

—HERMAN MELVILLE TO SARAH MOREWOOD

9

LEVIATHAN

Winter came quickly to Arrowhead and Broadhall in 1850, blanketing the farms in snow. By early December the snow was everywhere, the skies were gloomy, and a cold wind was sweeping across the fields. As the storms raged, Melville sat at his desk upstairs looking north toward Pittsfield, with Broadhall to his left, and Dr. Holmes's summer place to his right, its windows shuttered for the winter. Day after day, his pen raced across the pages of his manuscript, and the book that would become *Moby-Dick* began to assume its final shape. All the while that he was working at this high window, with the north winds rattling the glass, he had the strange feeling that he was at sea again. "I have a sort of sea-feeling here in the country, now that the ground is covered with snow," he remarked in a letter. "I look out of my window in the morning when I rise as I would out of a port-hole of a ship in the Atlantic. My room seems a ship's cabin; & at nights when I wake up & hear the wind shrieking, I almost fancy there is too much sail on the house, & I had better go on the roof & rig in the chimney."[1]

In the distance, clearly visible from his window, was a lumpish shape on the horizon that might have been mistaken for a leviathan coming up for air. The sloping form of Mount Greylock was about fifteen miles north of Arrowhead, and for those among his family and friends who knew the subject of his book, it was easy to imagine the author rubbing his eyes at twilight and wondering whether the great whale in his novel had escaped and was swimming across the hills.

For those who have fallen in love with the Berkshires at the height of summer, it is a sobering experience to know it in the depth of winter, when the snow hardens into crusty mounds, and the treetops that were so full and green are reduced to skeletal shapes against the sky. Even in midafternoon the light some days can take on the bluish tint of twilight and fool the eye into thinking the surrounding mountains are islands emerging from a sea. Over in Lenox, snowbound at times in his cottage, Hawthorne recognized the connection between Greylock and the Whale. "On the hither side of Pittsfield sits Herman Melville," he wrote, "shaping out the gigantic conception of his white whale, while the gigantic shape of Greylock looms upon him from his study window."[2]

Up by dawn each morning, Melville would spend a long day writing at his desk before taking a break when the sunlight at his window began to fade. He couldn't afford to waste any of those precious hours of winter light. While the sun tried to penetrate the gloom, he was determined not to do anything but write. It was a solitary life, with his family trying to stay out of the way as much as possible. His mother and sisters had their own domestic interests to occupy them, but Herman's punishing work schedule didn't seem as bad to his mother as his reluctance to trust his future to her Christian faith.

The unspoken burden of living with his mother was enduring her religious zeal. Though her daughters tended to follow her lead dutifully in most things, her son quietly insisted on doing things his way. It wasn't enough that he was providing a roof over her head. She cared little for his books, but professed to worry greatly over his soul. When she complained about him she often spoke as though no one in the family could match the ardor of her faith. As she confessed to one of her daughters, she longed to see the day "when all my belove'd children will in truth & sincerity openly come forward before the world & proclaim themselves on the side of God, feeling in their hearts their own unworthyness, trusting in the atonement of their blessed Redeemer Jesus Christ."[3]

From Thanksgiving to the end of the year, Lizzie and young Malcolm were absent from Pittsfield, spending part of the brutal winter at Judge Shaw's home on Beacon Hill, where the blasts of cold weather were easier to bear, so it was Maria Melville who ruled Arrowhead in that period. Whenever Herman left his upstairs room, he walked back into the same kind of life he had known since childhood, sitting beside the fire with his mother and sisters as though little had changed since he had gone to sea. Lizzie seemed content to have all that time away from Arrowhead, where she was never truly mistress of her own home as long as her mother-in-law was there. She later recalled that period as one of great strain and hardship. "Wrote White Whale or Moby Dick under unfavorable circumstances," she noted of her husband's work in her plain, telegraphic prose. "Would sit at his desk all day not eating any thing till four or five o clock—then ride to the village after dark—Would be up early and out walking before breakfast—sometimes splitting wood for exercise."[4]

He did look forward to his daily trips to the village in the family sleigh, bundling up for the ride in the heavy layers of "buffalo and wolf robes." By the end of each night he was exhausted. His eyes hurt and his head was throbbing. "My evenings I spend in a sort of mesmeric state in my room," he wrote to Evert Duyckinck, "not being able to read—only now & then skimming over some large-printed book." At first light, he repeated the same pattern of the day before. "I go to my work-room & light my fire—then spread my M.S.S. on the table—take one business squint at it, & fall to with a will."[5]

His steely resolve to complete the book by summer was fitting for a novel that is a kind of war story, a violent contest at sea between two implacable opponents. Ishmael certainly thinks he is at war. Even before the *Pequod* sets sail, he muses on the contradiction of the peace-loving Quaker owners of the ship sending it to invade the watery realms of the whale and spill great loads of "leviathan gore." In its size and power, the whale is like a "line-of-battle ship," and Melville describes whole fleets of the species moving through the seas in "martial columns" like a "grand armada." Likewise, the harpooners are cast as warriors engaged in "mortal combat," and the body of one—Queequeg—is so tattooed and scarred that he "seems to have been in a Thirty Years' War." Ahab plots the movements of these men like a military commander. They are the key lieutenants "in that grand order of battle in which Captain Ahab would presently marshal his forces to descend on the whales."[6]

When the novel begins, Ahab is already a wounded veteran of this war, having lost his leg to the white whale. In his eccentric speech, he says that he is now like a sailing vessel that has lost one of its masts. "It was Moby Dick that dismasted me," he tells his crew; "Moby Dick that brought me to this dead stump I stand on now." It is this grievance that prompts him to announce that he intends to pursue the whale to the ends of the earth. "I'll chase him round Good Hope, and round the Horn, and round the Norway Maelstrom, and round perdition's flames before I give him up." Just as wars are often seen as battles between good and evil, so Ahab turns his enemy—the white whale—into the embodiment of every sinful, despised thing on earth. "He piled upon the whale's white hump the sum of all the general rage and hate felt by his whole race from Adam down." He attributes to the creature an "intelligent malignity" that really belongs only to the race of Adam. Spiteful, sadistic killing is a specialty of humans, though they often find good reasons to deny it.[7]

As so many readers have recognized, it isn't the white whale alone that drives the madness. The whale is the illusion that grips the mind at the height of the fever. The whale is the thing that, beyond all reason, must be chased and caught. It is that fatal urge lurking in every heart to break the usual restraints of life and reach too far, to go over the edge and not come back. So, without leaving the Berkshires, Melville waged two major battles that winter. One was the imaginary conflict of Ahab and the Whale, and the other was the very real struggle to save his career.

TOSSED ON THE WAVES OF HIS SNOWY SEA, with Mount Greylock looming ahead, the author wrote all winter long, a ghostly figure in his study—except when venturing out in the night under his "buffalo and wolf robes" like a hunter. It was indeed hard sailing that winter as Melville led Ahab and the crew of the *Pequod* farther and farther from New England to the deepest waters of the Pacific. Strangely enough, earlier in the summer he had given Evert Duyckinck the impression that the story was "mostly done," and that it wasn't a dark tale at all. His friend came away with the idea that the new book would be "a romantic, fanciful & literal & most enjoyable presentment of the Whale Fishery." At that point, no one besides Melville could imagine just how dark and apocalyptic the author of the relatively lighthearted *Typee* could become. The summer had changed him, and his new book would reflect both his fresh surge of ambition and his fierce determination to succeed at any cost.[8]

Wanting the book to be overpowering at every level, he was able for the first time in his life to create something monumental. It was a big story about a long voyage, and the hunt for a massive creature without equal, but what the hard, lonely winter brought to it was a passionate intensity and wild abandon never seen in his work before. He wrote like a man possessed because now he was such a man, his whole future riding on every word he put to paper. However "fanciful" and light the story once was, the winter turned it into one of the great tales of dark obsession. There was no Ahab in his past, as far as we know, but all those grueling days at sea long ago on whaling ships engaged in a relentless slaughter had given Melville the kind of experience that could launch a thousand nightmares. Add to that his familiarity with the ghastly ordeal of the *Essex* disaster in the 1820s—with its terrifying details of a whale turning on its predators and sinking their ship, leaving the crew to the mercy of the waves—and Melville had all the material he needed to spin a tale that inhabits the border between the misery of madness and the delight of discovery.

What helps to set *Moby-Dick* apart from a mere tale of adventure at sea is the fact that Melville was not just telling a story. He came to inhabit it, speaking as if from within the tormented heart of Ahab or the half-bedazzled, half-bewildered mind of Ishmael. Melville is a Jonah riding not only in the whale, but also in the *Pequod*, the waves, the clouds, the passing birds, and the spirit of the universe itself. That is one reason it is sometimes difficult in the novel to tell who is speaking—Ishmael, Ahab, an all-seeing narrator, or one of the crew. Melville doesn't seem to care—all the voices carry something of his, embellishing the voyage with strange poetic outbursts, or interrupting it with philosophical musings and furious rants. Inspired as if for the first time, he created a narrative large enough to contain so many hopes, fears, and demons that readers of all types have been astonished to find something of themselves floating through the story, flashes of their own obsessions in watery mirrors.

Thanks in no small part to his romance with Sarah, Melville didn't need much in the solitude of his study to fire his imagination or his ambition. He was in the grip of his own obsession, and was speeding along a dangerous path with the same unswerving force that propels Ahab across oceans. His book and his debts had locked him into that path, and there was no going back. "Swerve me?" Ahab asks. "The path to my fixed purpose is laid with iron rails." Melville, at the helm of the imaginary ship that Arrowhead had sometimes become in his mind, was no less fixed in his purpose. His book, he would later say, was "broiled" in "hell-fire." What Ahab says of the power of his will, Melville himself could have echoed: "What I've dared, I've willed, and what I've willed, I'll do!"[9]

Ahab's madness is epic, but in the little world of Pittsfield, Melville must have seemed on the verge of madness himself that winter: spending money he didn't have, uprooting his family from New York for no logical reason, boasting of wild schemes to build a road and a tower in his woods, and locking himself away for the winter to write a book about "the Whale fishery." A good chief mate like Starbuck in the novel could have reasoned with him, but he wouldn't have listened to such talk any more than Ahab does. At the most basic level of *Moby-Dick*, all the mad pursuits at sea, the cries of the crew, the captain's diatribes, and the explosive violence of the great white whale bearing down on the *Pequod* are no more than phantoms in the mind of an author who fears that his brave but reckless pursuit of a dream will end in tragedy for all concerned. Yet he won't swerve to avoid the crash.

Looking back, Melville might have been tempted to follow this course regardless. He just needed the right push. The ex-castaway yearned to explore the edges of danger and test his limits. In that sense he identified not only with the hunters of whales and other objects of desire, but also with the great creatures whose dives in the unknown were spectacularly deep. "I love all men who *dive*," Melville famously said, and by that he was referring, of course, to anyone eager to plunge below the surface of things to discover what lies below the usual levels of perception. He made this comment just one year before starting *Moby-Dick*, explaining, "Any fish can swim near the surface, but it takes a great whale to go down stairs five miles or more; & if he don't attain the bottom, why all the lead in Galena can't fashion the plummet that will." Five miles was an exaggeration, but he liked the idea of the mighty whales exploring the depths—a mile or two down, at most—where no other mammal could hope to go. In part, *Moby-Dick* is the result of the author's own extended dive into the depths of his life. It allowed him to explore the mysteries of his identity, his dreams, and his experiences in new and complex ways. With Sarah and Arrowhead in the balance, he had every incentive to dive down as far as he could go.[10]

10

REVERIES

At Princeton University one night in November 1951, a Harvard professor told an audience celebrating the one hundredth anniversary of *Moby-Dick*'s publication that Melville's book was "wicked." Dr. Henry A. Murray wasn't a rabid moralist trying to ban the novel from college campuses, and he wasn't an old-fashioned English professor trying to exclude from the literary canon the upstart Melville, whose work was then enjoying its long-overdue acceptance by so much of the academic establishment. In fact, Dr. Murray lacked any literary credentials, but he was one of the most distinguished psychologists of his generation, a student of Carl Jung, and a major figure at the Harvard Psychological Clinic, with an M.D. as well as a Ph.D. When he called *Moby-Dick* wicked, he was praising the book, not condemning it, which must have made his comment seem all the more shocking to an audience in the 1950s. He insisted that any true appreciation of the novel had to take into account Melville's own remark when the work was nearly finished: "I have written a wicked book," the author had told Hawthorne, his fellow student of sin.[1]

The Harvard psychologist, who had also spent decades reading Melville and collecting information for a biography, was pretty sure he knew the itch the novelist was trying to scratch—the source of all the vitriolic attacks hurled against God and man in *Moby-Dick*. Drawing on his professional expertise, he deduced that it was a classic Freudian case: "When one finds deep-seated aggression . . . aggression of the sort that Melville voiced—one can safely attribute it to the frustration of Eros." Spirited and brilliant, the author of *Moby-Dick* must have been at war with his culture and its deity, Dr. Murray argued, because he hated the condemnation and guilt associated with pleasure and sexual freedom. Eros was seen as a force leading to "depravity," yet Melville yearned to experience it only as pleasure.

Murray even had a formula for the problem: "An insurgent Id in mortal conflict with an oppressive cultural Superego." It was a lot of intellectual weightlifting just to say that sex must be lurking somewhere in all the violent energy unleashed in *Moby-Dick*. Without knowing particularly why, many readers of the novel have felt that there is something sexual in the exuberance of Melville's prose, the lush, sensual quality of his descriptions, and the fierce intensity of his vision. The British writer Rebecca Stott has called it "the eroticized audacity of *Moby-Dick* . . . the briny adrenaline rush of its quest." Murray was on to something, and that's why it's odd that this towering figure in academic psychology discovered vital information about Sarah Morewood, yet saw nothing in it to reinforce his case. Summer after summer, with a real love for Melville's work, he traveled to Pittsfield searching for clues to various mysteries, helping to recover such treasures as the only surviving letters from Herman to Sarah.

(The Berkshire Athenaeum's Melville Room, where the letters are now accessible, "was established by the planning and generosity of Dr. Henry A Murray.")[2]

The Harvard professor quoted from one of these letters in his address at Princeton. He used it to explain that Melville wanted *Moby-Dick* to challenge readers and to make them reconsider their comfortable assumptions about life and art. What Herman told Sarah after he finished the novel was that she should "warn all gentle fastidious people from so much as peeping into the book." *Moby-Dick* was strong medicine and would overwhelm such people. "A Polar wind blows through it, & birds of prey hover over it," he said of the novel. It was probably this letter that led Murray—and others since—to discount Sarah's importance because, if Melville's words are taken literally, he seems to be telling this young wife that she is one of those weak, "fastidious" readers. Without any knowledge of her untamed character, or an awareness of the literary tastes she shared with Herman, it would be easy to mistake the extravagant, playful tone of his letter for something more serious:

> Dont you buy it—dont you read it, when it does come out, because it is by no means the sort of book for you. It is not a piece of fine feminine Spitalfields silk—but is of the horrible texture of a fabric that should be woven of ships' cables & hausers. A Polar wind blows through it, & birds of prey hover over it. Warn all gentle fastidious people from so much as peeping into the book—on risk of a lumbago & sciatics.[3]

This last joke suggesting that back pain will be the worst result of reading *Moby-Dick* is enough to give the game away. It's a very funny passage that uses self-deprecating wit to confide real truths about the book to a woman who knows the author better than anyone. It was always under the cover of language hidden in plain sight that these two lovers of words could safely communicate, knowing that no one else around them would catch the subtleties.

One of the inside jokes in this particular letter comes from the fact that it was partly in response to the gift of another book from Sarah. During her six months overseas with Rowland in 1850–51, she acquired a number of British books, and one of them was by a woman who was the exact opposite of the pampered lady toying with "fine feminine" silks. This author was Harriet Martineau, a tough, uncompromising journalist, lecturer, and novelist who established herself over a long career as one of her generation's leading feminists. Her most ambitious novel, *The Hour and the Man,* is a long piece of historical fiction about Toussaint L'Ouverture's slave revolt in Haiti, and it's far more violent than *Moby-Dick*. A scene in which children are torn to pieces by ravenous bloodhounds is especially gruesome. Yet this is the book that the supposedly "gentle" and "fastidious" Sarah bought, read, and then sent to Melville as a gift. She knew that its graphic descriptions of an island revolt might appeal to him, but he couldn't resist teasing her that his book had a texture that was even more "horrible."

Henry Murray didn't need Freud to understand Melville's yearning for a woman like Sarah. In a world full of proper ladies carefully brought up to focus their minds only on "gentle" subjects, here was a woman open to almost anything. You could even joke with her about masturbation, as a closer reading of Herman's letter might have shown Dr. Murray. But, first, it is useful to know a little bit about Pittsfield's most famous moralist in Melville's day—its most ardent defender of America's "oppressive cultural Superego," as Murray would have put it, and New England's preeminent authority on the wickedness of masturbation.

HAWTHORNE, HOLMES, AND MELVILLE PALED as cultural giants in Pittsfield when placed against the blazing star of Rev. John Todd, pastor of the First Congregational Church. His followers were in awe of him, all the clergymen in the nearby towns looked up to him, and thousands of schoolboys and college men tried to follow the moral code outlined in his bestselling *The Student's Manual*, which had been reprinted twenty-four times by the 1850s.

A Melville family copy of *The Student's Manual* survives. It belonged to Allan, Herman's younger brother, but almost every well-read young man knew the book, if only by reputation for its terrifying chapter 4, which was innocently titled "Reading." The page headings in Allan's copy—"A delicate subject," "Their certain ruin," etc.—highlight the real subject of the chapter. When young men read too much, Todd explained, and especially when they read the wrong books, their minds will begin to stray. The next thing you know they will be overcome by dangerous "reveries," their hands will become the devil's playthings, and then . . . The final step was so "delicate" that Todd had to switch to Latin to explain the worst details, but a footnote made sure that any boy whose Latin was insufficient could see the ruin awaiting him in a moment of weakness. The act was "very frequently the cause of sudden death," Todd warned. You could be so immersed in the sinful deed that you wouldn't even realize God was just waiting to fell you with a stroke. "The apoplexy," the reverend darkly informed his readers, "waits hard by, as God's executioner, upon this sin."[4]

Because Todd couldn't bring himself to name the dreaded act in English, he relied primarily on "reverie" as his preferred euphemism. It was bad enough for boys to allow their suspect reading to lead them into a reverie, but Todd thought that some of the most sinful creatures were the authors of dangerous books whose words tainted the minds of innocent youth. These "bad" authors, of whom he could bear to name only half a dozen, including Lord Byron and Edward Bulwer-Lytton, were doomed to suffer unimaginable torments. "They dig graves so deep that they reach into hell."

The trembling youth anxiously turning the pages of Todd's chapter on reading found only two ways to avoid joining Byron and Bulwer-Lytton in hell. First, avoid at all costs "the habit of reverie"; second, do not read bad books—"I do entreat my young readers never to look at one—never to open one." These are the words that Melville is parodying when he writes of *Moby-Dick* to Sarah, urging her in a mock tone, "Dont you buy it—dont you read it." His letter prepares her for the joke a few sentences earlier by a sly remark about their shared love of "falling into the reveries" of their books. "A fine book is a sort of revery to us—is it not?" he adds coyly.

Sarah knew very well what he meant. She read a famous rejoinder to Todd's obsessive fears—*Reveries of a Bachelor*—around the time it first appeared in early 1851. The author, Donald G. Mitchell (or Ik Marvel, as he inexplicably called himself), invited his readers to go ahead and indulge their imaginations and not restrain themselves. "And have you not the whole skein of your heart-life in your own fingers to wind, or unwind, in what shape you please? Shake it, or twine it, or tangle it, by the light of your fire, as you fancy best." Likewise, in his own "wicked" book, Melville took direct aim at Rev. Todd and the other self-appointed guardians of virtue for turning pleasure into their idea of "depravity." In the chapter of *Moby-Dick* called "A Squeeze of the Hand," Ishmael falls into reveries by the score as he joins his fellow sailors of the *Pequod* to squeeze a tub full of lumpy whale spermaceti into liquid.[5]

> I squeezed that sperm till I myself almost melted into it; I squeezed that sperm till a strange sort of insanity came over me. . . . Come; let us squeeze hands all round; nay, let us all squeeze ourselves into each other; let us squeeze ourselves universally into the very milk and sperm of kindness. Would that I could keep squeezing that sperm for ever!

Modern readers get the point, and some have even interpreted this passage as an example of homoeroticism in Melville's work. He was, in fact, always fascinated by the possibilities of male friendship, and his openness to the idea of intimacy between men is obvious in his treatment of the warm relationship between Ishmael and Queequeg. Yet part of the fun of sharing jokes with Sarah was that she was liberated enough to entertain all sorts of forbidden thoughts. Very little seems to have shocked her. It's no wonder that so many of Melville's contemporaries were left bewildered by the strange fun the novelist was having in the middle of such a grim tale of obsession. Apart from Sarah, few readers in the 1850s could appreciate even half of what Melville was trying to say about the battle Dr. Murray described in the 1950s as the id combating the massive, whale-like bulk of the "cultural Superego."

Whatever name it's given, the battle was staged in great part for Sarah's benefit—to amaze her, amuse her, and to conquer the world for her. At the most basic level Melville needed to conquer the book trade first, enabling him to be independent and to sustain a lifestyle that Rowland Morewood could afford but that he could not. All the same, the author of *Moby-Dick* is chasing a dream that is much grander than dollars, though the dollars will be necessary to him in the end, which is the inescapable reality he must face as he struggles to make a whaling yarn sing like poetry. Fortunately for posterity, poetry prevailed, allowing reverie to triumph over reality.

As Melville sensed at the beginning of his work on the story, the great artistic challenge he faced was to find a poem of life in something as elemental as "blubber." Just before his summer at Broadhall, he told fellow author Richard Henry Dana Jr. in May 1850,

> It will be a strange sort of a book, tho', I fear; blubber is blubber you know; tho' you may get oil out of it, the poetry runs as hard as sap from a frozen maple tree; —& to cook the thing up, one must needs throw in a little fancy, which from the nature of the thing, must be ungainly as the gambols of the whales themselves. Yet I mean to give the truth of the thing, spite of this.[6]

Though no woman plays any part in the major action of *Moby-Dick*, its muse was Mrs. Morewood. She was the ghost that Dr. Murray almost glimpsed between the pages when he saw "the frustration of Eros" looming so large over the masterpiece he loved. It was her spirit that fueled Melville's dreams for a different kind of life, opening that hidden vein of poetry which runs so wildly through *Moby-Dick*.

11

BLACK QUAKE

In the rolling hills of Derbyshire—just west of one of England's grandest houses, Chatsworth—the Morewood family had their own modest estate on land they had owned since the eighteenth century. Their handsome old house was called Thornbridge Hall, and it looked out on a green and pleasant land where the family had long prospered. Over several generations there had been at least four high sheriffs of the county who were Morewoods. The current patriarch—Rowland's father, George—had been born in 1763, and was still working hard to expand his fortune in his eighties.

Though Thornbridge Hall overlooked some of the prettiest scenery in the north of England, Sarah would have found little there to keep her entertained except her books. Business always came first in the Morewood family—at least in those days, when they had such a wide and diverse range of interests. They were not the sort to dash off at a moment's notice for picnics and long sightseeing tours. Travel was for making money. Most of their early fortune came from the cotton trade, and at one time they had offices not only in England and America, but also in Russia. It is unlikely that Sarah could have endured her long stay in the lonely isolation of Derbyshire, and at another family home in Lancashire, without writing to Melville, but none of her letters to him have been found. "Make a fire bright with my letters and oblige me," she once told another man to whom she sent flirtatious messages, and no doubt she advised Melville to do the same.[1]

Throughout the long winter of her absence, Melville certainly made a point of picking up his own mail in town, not leaving the task to anyone else in his household. For the first three months he also always insisted on driving his mother, sisters, and wife everywhere in the family wagon or sleigh, even when it took precious time away from his writing. It would seem he didn't want to risk having a letter from England arrive and fall into the wrong hands. Sarah had a real talent for writing letters, and it is a great loss that we have none to Melville. Her casual prose outshines much of her poetry, for it tends to be more direct and vivid. Writing to one of her friends in New York, she described a winter day in the Berkshires as having "a mantle of snow wrapping the hills in a shroud—as it were—taking from them the friendly look they used to wear." The "bleak north wind" was so strong, "the shutters of our house refuse to remain fastened . . . and bang as if at war with their hinges,—a dismal echo through empty rooms."[2]

There were acceptable reasons for Melville to keep in touch with the Morewoods. His cousin spent the winter at Broadhall acting as caretaker, and Herman could have sent Sarah a few seemingly innocent updates about the house or the farm, relying on their usual literary references to slip more important information into the letters. One sad event at Broadhall that winter almost demanded that he write to Sarah. It was news about the three horses she left behind to await her return in the spring. They had been spooked by a train that ran near the edge of the pasture, and one of them—Black Quake—had raced into the locomotive's path.

Melville blamed his cousin Robert for the horse's death. Robert had been driving a sleigh toward the tracks, with the three horses following close behind, and when he cracked a whip to speed over the rail crossing, they ran after him. Black Quake broke "his leg clean into two pieces," Melville learned afterward, and didn't survive. The other two were shaken, but would be all right. Knowing the strength of Sarah's bond with her young horse, Melville was the one person who could send her the news in a way that would soften the blow. But simply in recalling the incident in a letter that has come down to us intact—once again to Evert Duyckinck, who seems to have saved everything—Melville began to conjure fond images of Sarah and her horse, recalling that memorable weekend of the costume party when she dressed as saucy Aunt Tabitha, and he was a bride-abducting Turk. That was also the weekend when Sarah rode Black Quake into the dark green wilderness of the Gulf Road—Saturday, August 10.

Soon he was lost in this summer memory. The more he thought about that colt, with its "bounding spirit and full-blooded life," the more he identified with it. At that point the local farmers had not yet given up hope that Black Quake could be saved, so what came to Melville was a vivid image of the lame colt no longer able to bear "Mrs Morewood on his back." Suddenly, he realized that the real loss here was one he could understand. He said that what had happened to the colt was "not one jot less bad than it would be for me." In other words, as he now saw it, the tragedy for the horse was not merely the broken leg, but the loss of all those future summers in Sarah's company. For a moment he saw himself and the horse as one, and he was sorry to think how that creature "might for many a summer have sported in pastures of red clover & gone cantering merrily along the 'Gulf Road' with a sprightly Mrs Morewood on his back, patting his neck & lovingly talking to him—considering all this, I say, I really think that a broken leg for him is not one jot less bad than it would be for me."[3]

Here, almost as a coy, throwaway line, Melville was openly admitting that a future without Sarah close to him would be heartbreaking. Duyckinck might have thought he was making some joke about broken legs, but that wasn't the "bad" thing he was "considering." It was the prospect of losing a long idyllic future in summer pastures at Broadhall, with all those merry rides into the lush countryside, and all the other pleasures he was now missing on a snowy December evening, writing by candlelight, and listening to the wind howl. "Not one jot less bad than it would be for me" was a convoluted and whimsical way of suggesting, "I miss Sarah, and I miss her so much that I'm even going to share my secret tonight with you, Evert Duyckinck, if you pause long enough to think about it."

This was Melville pointing from the date of his letter, December 13, 1850, backward to August 10, 1850, and then forward to the coming summer of 1851, and imagining how bad that season would be if Sarah didn't return, or if he failed to keep his tenuous hold on Arrowhead and wasn't waiting for her. (He was in a confessional mood that night, volunteering a detailed look at his typical day with the unprovoked question "Do you want to know how I pass my time?")

THREE DAYS LATER, on December 16, Melville did something unexpected while he was writing *Moby-Dick*. He stopped the narrative and put himself, along with his moment in time, straight into the passage he was working on. He was in the middle of one of his pedantic chapters, debating whether whales spout water or just vapor, when he suddenly removed the mask of the novelist and stepped forward to announce that he was writing his book at "fifteen and a quarter minutes past one o'clock P.M. of this sixteenth day of December, A.D. 1850." What was the point of doing this if not to remind readers that behind all the impersonal conventions of fiction there is always a real writer looking back at you with bills to pay, dreams to pursue, sorrows to bear? There is no way to know exactly what Melville had on his mind at one that afternoon, but what if the thought wasn't as important as the act of marking the date, as a prisoner might do in a cell for the time of release?[4]

He was in the grip of something that had to conclude soon—for good or ill. He was on a long, hard voyage to that conclusion, sticking to a grueling schedule that he felt compelled to share with Evert only three days earlier. On reflection, "Do you want to know how I pass my time?" does indeed sound like a question from an inmate trying to fill the days until the sentence is served. It's what might be called the log mentality, natural enough to sailors keeping track of a voyage, but also useful for prisoners. In that study at Arrowhead where he locked himself away for much of the winter, Melville was both a mariner and a prisoner.

Marking the date also highlighted for Melville a one-year anniversary of special interest now that Sarah was abroad. The previous year at this time, he had been visiting the same country where she was now. For most of November and December 1849, he was in London dealing with publishers and searching for inspiration for his next book. Toward the very end of his stay, he seems to have found the spark for his story of "the Whale fisheries."

What he was doing now at Arrowhead began the previous December when the work of the greatest English painter of the nineteenth century revealed to him the artistic potential of the mighty leviathan.

12

THE VORTEX

In a bookcase of the Melville Room at the Berkshire Athenaeum is a large blue volume of engravings that once belonged to Sarah Morewood. In the Pittsfield of her day, *Turner's Rivers of France* would have stood out immediately as an unusual and expensive work to grace any shelf. J. M. W. Turner's genius wasn't so widely recognized in those days, but there was at least one local connection to this English painter. Herman Melville was such an enthusiastic admirer that his own treasured collection of engravings would eventually include at least thirty-three from Turner's work. While Melville was writing *Moby-Dick*, there were some in the English press who thought young Melville and Turner were kindred talents.

As the London *Athenaeum* argued in 1850, Turner and Melville had no other peers when it came to capturing "the poetry of the Ship—her voyages and her crew." Much of Turner's work was sure to appeal to Melville because it featured so often the majesty of the sea, and the stark beauty of the tall ships. He couldn't help but iden-tify with a great artist boldly willing to impose his vision on a scene that others saw only in the most literal terms. Turner's work showed him how to take real experiences at sea and merge them with the swirling impressions of something greater and more imaginative.[1]

Turner's fondness for experimenting with colors and shapes cre-ated mysteries in his canvases that realists abhorred. This was es-pecially true in the 1840s when Turner took up a subject that was entirely new to him. A rich patron—Elhanan Bicknell, an English-man in the whale oil business—encouraged him to create a few works featuring whaling ships and their prey. The painter had never been at sea in one of the ships, but the possibilities intrigued him, especially because most people knew so little about the look and movements of these enormous sea creatures.

Bicknell gave Turner a learned book on the subject—Thomas Beale's *The Natural History of the Sperm Whale*—and after carefully examining various sections of it, Turner set to work. ("Turner's pic-tures of whalers were suggested by this book," Melville wrote on the title page of Beale's history when he bought a copy in 1850.) Neither the businessman nor the critics liked the results of the painter's ef-forts. Fashionable London wasn't ready for Turner's whales. Critics ridiculed two of his paintings that were exhibited at the Royal Acad-emy. One of the comic wits of *Punch* suggested that one of the paint-ings was really a shadowy sketch of "lobster salads." More serious was the attack in William Harrison Ainsworth's *New Monthly Mag-azine,* which claimed that Turner's pictures were so garishly ugly that they cast a shadow over every other work exhibited at the Royal Academy. "Mr. Turner is a dangerous man," warned the magazine, "and ought to be suppressed. But if he must continue to work in this brimstone vein, he ought to have a small apartment to himself, where he could do no harm."[2]

The common complaint against Turner's work of this period was that his views of real objects were "indistinct," and no one could tell what was what. The bold use of color, the dreamlike visions, the subtle play of light, and the mere hint of distant shapes were annoying to buyers accustomed to lush realism. When an American collector complained about one of Turner's paintings being "indistinct," the artist took the criticism in stride, saying, "You should tell him that indistinctness is my fault." When the next exhibition at the Royal Academy opened, Turner defiantly came back with two more paintings of whaling ships—one of which bore the clumsy and unattractive title *Whalers (Boiling Blubber) Entangled in Flaw Ice, Endeavouring to Extricate Themselves*. He was thumbing his nose at the marketplace, brazenly showing his independence and his refusal to bend to its demands. Of course, both went unsold, but so too did one of the great masterpieces of the century, his apocalyptic vision in a brilliant gold tint, *The Angel Standing in the Sun*—a disturbingly beautiful explosion of light at the end of time.[3]

Like Turner at his easel, Melville learned to make a virtue of the fault of indistinctness. *Moby-Dick* is the literary equivalent of a gallery filled with the best of Turner's canvases. So much of the book shows the painter's influence, which can be felt in the bold sweep of the story, in the iridescence of the language, and in the author's frequent willingness to cast a suggestive haze over certain scenes. *Moby-Dick* features an overwhelming collection of powerful scenes in which the shapes we know from reality float in a tumultuous wash of colors and images spilling from the artist's eye. The effect is what happens when the "great flood-gates of the wonder-world"—as Ishmael calls them—swing wide, producing a cascade of suggestive impressions whose full force may not be understood for years. The first image of the great white whale in *Moby-Dick* is a Turner oil in eleven words— "one grand hooded phantom, like a snow hill in the air." (It was also a good description of Mount Greylock that winter.)[4]

When Ishmael is in New Bedford at the beginning of the book and decides to visit the Whaleman's Chapel, a painting catches his eye as he watches Father Mapple mount the pulpit to deliver a sermon on Jonah and the whale. A large work dominating the wall behind the preacher, it shows a ship in distress, fighting to stay afloat in a violent storm "off a lee coast" where the winds have brought it dangerously close to "black rocks." But in the bleak sky overhead there is a bright beam of sunlight breaking through, and in the middle of that radiance is an angel offering hope—or so Ishmael believes. He thinks the angel is saying, "Beat on, thou noble ship . . . for lo! the sun is breaking through; the clouds are rolling off—serenest azure is at hand."[5]

In a book that consistently questions conventional religious dogma, this vision of an angel emerging from the sun is not some pious appeal to faith, as young Ishmael—yet untested by his voyage on the *Pequod*—innocently presumes. It is a profound omen, a sign of the dangers awaiting Ahab and his crew when they commence their voyage on Christmas. As a warning, it is as wasted as all the others because the temptation to overreach for glory or revenge or pride is so strong. But there is something much darker at work here. It is one of the first hints in the book of Melville's fear that all human endeavor, no matter how grand or seemingly righteous, is doomed to fail and be tainted by what he calls the "horrible vulturism of earth"—the universal predatory urge that sweeps humanity into countless voyages to chase and capture one thing or another. It is the vision that Turner brought to his canvas when he painted the work that Melville alludes to in the description of the chapel painting—*The Angel Standing in the Sun.*[6]

In Turner's masterpiece the angel standing in the middle of the sun may look benign at first glance, but it is the archangel Michael holding his sword aloft on the Day of Judgment, and in the molten gold of the light pouring down from him, there is nothing below but a few frightened humans fleeing in panic and despair. Overhead, like a blot on the canvas, is a swarm of black birds circling in a frenzy. It was one of the great magical effects of Turner's brushwork that he was able to create the impression of an overwhelming light circling outward and turning inward at the same time. Light flows from the angel while seeming to collapse into a vortex at the outer edges, swallowing the shadowy humans below. Then again, Turner was a master at painting an elemental vortex, especially at sea. What the noted painter and art critic Sir Lawrence Gowing says of light in Turner could also be said of humanity in Melville's vision of a universal vulturism: "Light is not only glorious and sacred, it is voracious, carnivorous, unsparing. It devours impartially, without distinction, the whole living world."[7]

Near the end of *Moby-Dick,* as Ahab closes in on his fatal encounter with the whale in what he calls his forty years of "war on the horrors of the deep," the dawn comes up one morning in a blaze of light that drenches the whole scene and turns the sea into "a crucible of molten gold, that bubblingly leaps with light and heat." It is a frightening but gorgeous Turneresque moment, with sunlight flashing in all directions. Blind to the approaching danger, Ahab mistakes this explosion of light as nature yielding to his will. At this moment, he thinks he is not only a great captain, but a Neptune whose ship is a "sea-chariot" towing the sun across the world, independent of any other power on earth. Fueled by his "fatal pride," Ahab soon finds his whale, and the two mighty enemies engage in a battle to the death that will sink the *Pequod* and send every member of its crew to the bottom—except Ishmael. After the whale rams it, the ship is swallowed in a ravenous "vortex" of water. The famous scene is nothing less than a prose version of Turner's *The Angel Standing in the Sun*, complete with "archangelic shrieks," as Melville puts it, accompanying the disappearance of the mainmast beneath the waves, and the flocks of birds suddenly appearing overhead: "Now small fowls flew screaming over the yet yawning gulf."[8]

HOW DID A YOUNG WRITER in Pittsfield learn so much from a painter he never met in a country thousands of miles away? Periodicals, books, and museum visits in London taught Melville some of what he knew about Turner. In December 1849, he spent parts of two days that month with one of Turner's oldest friends, the poet and wealthy collector Samuel Rogers. In the Royal Academy catalog for *The Angel Standing in the Sun*, Turner cited two lines from Rogers as the only contemporary inspiration for his painting: "The morning march that flashes to the sun; / The feast of vultures when the day is done."[9]

Thanks largely to the high reputation of *Typee* in England, Rogers welcomed his young American visitor warmly, and gave him a tour of his spacious townhouse in London, near Buckingham Palace. It was as much a museum as a residence, with the walls adorned by some of the finest art in any private home—a Rembrandt self-portrait, a large historical scene by Rubens, a majestic Titian, four paintings by Thomas Gainsborough, five by Nicolas Poussin, eleven by Sir Joshua Reynolds, and several works by Turner. On December 20 Melville stayed for three hours, and came back on the twenty-third for another three hours with Rogers, who was a walking encyclopedia of British life and culture, an old man famed for his anecdotes about the many famous people he had known—from Romantics like Byron to the recent Victorians like Tennyson. There were few men of genius he knew better than Turner, who was then ill and nearing the end of his life.

Samuel Rogers and Turner had worked closely together in 1830 on an expensive book of poetry and engravings called simply *Italy*. ("An interesting book to every person of taste," Melville would later say of it.) A few years later the two men combined their talents again to create an illustrated version of one of Rogers's best poems, *The Voyage of Columbus*. The poet conceived of the explorer as a man of peace betrayed by future generations who would fight over the New World and spoil paradise with the violence and greed of the Old. It was in this poem—in a canto titled "The flight of an Angel of Darkness"—that Rogers so neatly captured the disillusion of heroes who begin their adventures with high hopes ("The morning march that flashes to the sun") only to find death and defeat in the end ("The feast of vultures when the day is done").[10]

For two mornings in London, Herman Melville was the guest in a house where poetry, painting, angels, vultures, Turner's career, fame, and ambition could all occupy the same intellectual space in a setting that was a veritable temple of art. The young man would never again be entertained in any place that could compare with Rogers's house at 22 St. James's Place. An engraving by Charles Mottram in 1815 shows Byron, Coleridge, Sir Walter Scott, Turner, and Wordsworth gathered around the very same dining table where Melville was served breakfast in December. This was a heady moment for an American writer with so few years of literary experience behind him. It provided the kind of inspiration that Melville would describe in his Hawthorne essay: "Genius, all over the world, stands hand in hand, and one shock of recognition runs the whole circle round."

The treasures inside Rogers's home were worth far more than the building itself, but Rogers liked to think of his tall townhouse "as a fitting frame to a beautiful picture, or a precious binding to a rare book." More than his poetry or his money, this house had brought him the greatest renown in his old age. So many visitors to England longed for a glimpse inside that a London guidebook included a detailed list of its most valuable contents, and then added the discouraging note: "Mode of Admission—A letter of introduction (the only mode)." The information was in a section grandly labeled "Houses of the Principal Nobility and Gentry." (Besides his just claims as a celebrated author, Melville was also able to supply Rogers with a letter of introduction from Edward Everett, then president of Harvard and a former ambassador to Britain, as well as a friend of Judge Shaw.)[11]

AFTER LEAVING THIS HOUSE of artistic wonders and being regaled for hours with stories of great writers and great painters, Melville must have still had his head in the clouds when he sailed for America a few days later. Within weeks of his return to New York, he was hard at work on a story of a "whaling voyage," and joking with Richard Henry Dana that he was trying to find poetry in "blubber."

By the end of the summer, he had found some of that poetry, and he had found love. And now at the end of 1850—looking back at how far he had come since the previous December with Samuel Rogers—he was wondering whether his own progress from brilliant beginnings would end well, or turn into a disaster. He understood both the plight of the everyman caught in the larger designs of others and the exhilaration of the extraordinary hero aspiring to the highest pinnacle. He could write convincingly of both positions, and warn of the dangers facing both. But could he save himself? Or would his failure simply provide another "feast for vultures"?

13

THE ELUSIVE NEIGHBOR

In that long winter in the Berkshires, Melville was not the only one in the neighborhood writing an American classic. Over in Lenox, Hawthorne was working on his novel *The House of the Seven Gables,* and he was writing at a Melvillean pace. He began the book in August 1850 and finished it at the end of January. Unlike Melville's marathon with *Moby-Dick,* this literary sprint of Hawthorne's was a relatively painless endeavor. In their snug little house Sophia Hawthorne helped to keep everything running smoothly while her husband wrote in the mornings, and when he appeared at lunch, she and their two small children would welcome him like a hero, with "great rejoicing throughout his kingdom," as Sophia put it.[1]

Preoccupied with their books, Hawthorne and Melville saw each other infrequently during the winter. In January, when Melville went to Lenox for a brief visit, he was happy to see that his fellow author was doing well after a storm. "I found him, of course, buried in snow," Melville wrote to Duyckinck, "& the delightful scenery about him, all wrapped up & tucked away under a napkin, as it were." After a meal of cold chicken, Melville returned home, but with a promise that Hawthorne would soon visit him at Arrowhead. He was hungry for companionship and yearned to discuss his novel with the older writer over "a bottle of brandy & cigars." Without Sarah in the neighborhood, he had gone far too long in solitude thinking about art and life.[2]

When Hawthorne kept putting off the promised visit because of his work, the weather, and other concerns, Melville grew increasingly impatient. Here he was with those six miles separating him from a great American writer with insights into fiction, fame, darkness, and sin, yet weeks were going by without a word between them. He tried to make it clear that there was some urgency on his part. He wrote, half in jest, "Come—no nonsence. If you don't—I will send the Constables after you." A room at Arrowhead was waiting: "Your bed is already made, & the wood marked for your fire."[3]

At last, in mid-March Hawthorne came to Arrowhead for a short stay. The weather was still cold and raw, so the two novelists didn't venture outdoors much, but took shelter in the barn, where they could talk and smoke to their heart's content. Hawthorne found a carpenter's bench to be the best resting place, and there he would sit or lie for hours listening to Melville unload all the many weeks of thoughts he had been holding inside. Occasionally they stepped outside for air, to stretch their legs, and admire "a fine snow-covered prospect of Greylock." After a couple of days of indulging Melville's need for a sounding board, Hawthorne felt as if he had been at Arrowhead for a week. Recalling Henry David Thoreau's recent book, *A Week on the Concord and Merrimack Rivers*, Hawthorne joked as he was leaving that he should follow *The House of the Seven Gables* with a new volume called "A Week on a Work-Bench in a Barn."[4]

His dry wit didn't offend Melville, who would have kept him as a captive audience for a week if he could have. It was an old habit from his naval days to gather with friends on the foretop of the old warship *United States* and talk for hours. On dry land in the Berkshires, the carpenter's bench was the best substitute available for the airy platform high above the ship's decks. As he recalled in *White-Jacket*, "the tops of a frigate are quite spacious and cosy. They are railed in behind so as to form a kind of balcony, very pleasant of a tropical night. From twenty to thirty loungers may agreeably recline there, cushioning themselves on old sails and jackets." In his imagination, Melville must have spent many a day that winter in the foretop, planning and writing his novel as if from a great height, looking down on the *Pequod* and gazing far away at the sea as his cast of characters sailed along. In truth, it was such a lonely voyage that when Hawthorne finally came to spend a little time with him, he acted like a man who hadn't seen another soul for months.[5]

There were things about his work that he felt Hawthorne could understand especially well, and in an ideal world, Melville could see himself and his friend as brothers living and working side by side, engaged in endless speculations about the universe. In a burst of enthusiasm one day that spring, he wrote to Hawthorne to share his vision of a tropical eternity in which they would drink champagne and talk "in some little shady corner" until the earth is just "a reminiscence, yea, its final dissolution an antiquity." Such talk was simply too much for a private man like Hawthorne, and in the coming months he began finding more excuses to keep Melville at a comfortable distance. He didn't crave literary companionship of the kind the younger author sought. An intense and revealing relationship with another writer was the last thing he needed, and it was too much to expect an author of his stature to be anyone else's sounding board. Nevertheless, that was what Melville tended to want from him. "I know little about you," wrote Herman, "but something about myself. So I write about myself. . . . Don't trouble yourself about talking. I will do all the writing and visiting and talking myself." Soon even Melville feared that he had treated his friend too often as a choir expected to sit patiently through another long sermon. In one letter he apologized to Hawthorne: "I am falling into my old foible—preaching."[6]

It wasn't really in Melville's nature to bombard others with his thoughts or to seek closer bonds with writers generally. He was almost as private as Hawthorne, and sometimes even as reserved, but when he found the right person who seemed to share his views or sympathize with his aims, he came to life in an explosion of feeling. Once he opened his heart to someone, the force of his personality could be overwhelming. Sophia Hawthorne recognized this and responded to it better than did her husband. "He is an incalculable person," she wrote of Melville, "full of daring & questions, & with all momentous considerations afloat in the crucible of his mind. He tosses them in, & heats his furnace sevenfold & burns & stirs, & waits for the crystalization with a royal indifference as to what may turn up, only eager for truth, without previous prejudice."[7]

In time, Melville came to accept a difficult truth. Much as he admired Hawthorne, their temperaments were too disparate. The author of *The House of the Seven Gables* was not one who shared Melville's notion of diving deep. He was never going to join him in a fearless plunge into the most dangerous waters of the soul. He was always going to keep his head above water. So restrained and aloof was Hawthorne that Sophia once said of her husband, "He hates to be touched any more than anyone I ever knew." As a literary craftsman, he was the jeweler working in the quiet back room, while Melville was the sculptor dangling from the side of a massive stone.[8]

Perhaps one reason the sculptor interested Hawthorne is that, on occasion, he wanted to throw caution to the winds himself and take greater risks. A scene in *The House of the Seven Gables* suggests as much. It takes place at the long window in the old house when a noisy political parade passes by in the street below. Poor Clifford, the sad wreck of a man whom life has treated so unfairly, stands at the window in an agitated state and almost jumps into the middle of the crowd, but his sister and their young cousin Phoebe restrain him. Sounding very much like Melville, Hawthorne says in his narrative voice that Clifford might have been better off to jump. What he has in mind is more in keeping with Melville's metaphorical diving than the real thing: "He needed a shock; or perhaps he required to take a deep, deep plunge into the ocean of human life, and to sink down and be covered by its profoundness, and then to emerge, sobered, invigorated, restored to the world and to himself." But, of course, Clifford does not take that plunge. Such dives belong in Melville's work, not in *The House of the Seven Gables,* and the best Hawthorne can offer is the recognition that—for those brave enough, and reckless enough—the result might be worth the risk. Then again, Hawthorne adds with his typical good sense, it might simply result in "the great final remedy—death!"[9]

WHEN, IN APRIL, MELVILLE WROTE to Hawthorne about *The House of the Seven Gables,* which had just been published, he began by praising the book generally, then singled out Clifford's aborted jump as one of the best scenes in the book. It struck a chord with him, as Hawthorne must have known it would. Melville didn't try to interpret it or turn it into a commentary on Hawthorne himself. The book was selling well and getting good reviews, so there wasn't much that Melville could add—except to seize the chance to plead once again for a greater friendship between them. Perhaps thinking of his long separation from the woman he loved, Melville was struck by the fact that his elusive neighbor was so aloof that he might as well be in England. He felt it was necessary to remind Hawthorne that "the architect of the Gables resides only six miles off, and not three thousand miles away, in England, say."[10]

Such reminders had little effect. Hawthorne was not going to be William Wordsworth to Melville's Samuel Taylor Coleridge in an American Lake District. He was already thinking that a year in the Berkshires was too much. He was anxious to move. He missed the sea, didn't care much for the snow, hated the bitter cold, and no doubt believed that six miles was too close for comfort with a demanding friend like Melville, who was never subtle in his approach. Melville ended his letter about Hawthorne's new novel with another blunt order: "Walk down one of these mornings and see me. No nonsense; come."[11]

It took a while, but Melville began to get the hint that his fellow author preferred to be friendly at a distance. At first he tried to cover his disappointment by pretending that Hawthorne wasn't the only one who could stand aloof. Suddenly he was too busy and too tired to travel even the short distance to Lenox. "I feel completely done up, as the phrase is," wrote Melville, "and incapable of the long jolting to get to your house and back." Soon he changed his tone again, going from feigned weariness to wounded indifference. "Come and spend a day here, if you can and want to; if not, stay in Lenox, and God give you long life." This strain between the two doesn't fit the usual narrative of their friendship as something so close it was like a love affair. There was a good reason why Melville was finally able to tell Hawthorne to come or to forget it. By the time he wrote that remark, Sarah Morewood had come home from England. The real love affair in his life could finally resume where it had left off.[12]

IT WOULD TAKE the rest of the spring and part of the summer to finish *The Whale,* as he was then calling his novel, and he was feeling the pressure not only to get the book out, but to collect some money for his labors. It wasn't going to do any good to have Sarah near him again if he couldn't afford the financial burden of being her neighbor. On April 25, 1851, one week before she returned to America, he wrote to his publisher in New York—Harper & Brothers—asking for an advance on the new work. Lacking any strong sense of the book's prospects, the firm said no, causing a desperate Melville to make his problems worse by going even deeper into debt. On May 1, he quietly borrowed $2,050 from an old family friend, neglecting to tell his wife or father-in-law about it.

Melville spent part of the money on improvements to Arrowhead. A few changes to the house were necessary, but at least one was totally to please his fancy. Unable to afford his tower with a grand lookout toward Broadhall and Greylock, he paid instead to have a covered porch built at the side of his house with a view in the same direction. It would at least give him a place to sit when the weather turned warmer, though some visitors would wonder why he built it facing north. If he couldn't have a castle top for his perch, Melville could at least fancy himself on some Italian hillside watching the clouds drift by from a place he insisted on calling his "piazza." A modern visitor can sit there in shadows now and question why anyone would call it by such a name. It's just a porch, with a sloping roof, a narrow floor, and a few wooden steps leading up to it, but Melville was making an effort to cast a more romantic light over his poor substitute for Broadhall, his only foothold in the neighborhood.

In the long period of Sarah's absence, Melville's dreams of her return didn't keep him from taking some comfort in Lizzie's arms. However unhappy their relationship, and however unromantic their overcrowded house, they continued having sex, and—as winter came to a close—Lizzie discovered she was pregnant. For months afterward, the family spoke little of it. With so much debt hanging over him, Melville couldn't have been overjoyed to have another mouth to feed. "Dollars damn me," he told Hawthorne near the end of his work on *Moby-Dick*. "The malicious Devil is forever grinning in upon me, holding the door ajar." Like Ahab, he was in a race, impatient to overcome all obstacles and capture the greatest prize. What he was after would prove every bit as elusive as the white whale.[13]

14

TO GREYLOCK

The picnics, costume parties, fishing trips, and galloping rides over the countryside began again in earnest soon after Sarah returned to Broadhall in 1851. For a few weeks in the late spring she was in and out of New York, recovering from a stormy return voyage on the Atlantic in April, and organizing her move to the mansion that now sat vacant awaiting her arrival. Thirty-five years after Uncle Thomas had settled at Broadhall, the last of Melville's cousins had left the place and turned over the keys.

For the six months of Sarah's absence, Melville had stayed close to his desk and rarely ventured beyond Pittsfield. When spring arrived, he felt like a creature who had been in hibernation or one who came to life only after dark. His eyes were so strained from overwork that he seemed to be squinting at everything. "Like an owl," he wrote, "I steal abroad by twilight, owing to the twilight of my eyes." With the news of Sarah's return to America in the first week of May, he came to life and announced to his family that he must make a quick trip to New York. He was still trying to finish his novel, and had half a dozen chores to occupy him at Arrowhead, yet he dropped everything and raced away on a mad dash to the city by train, returning three days later. It's possible that he wanted to drop off part of his manuscript so that the long printing process could begin, but that simple task didn't demand such haste and expense. Whatever excuse he offered his family, it seems more likely that Sarah's presence in New York prompted this abrupt break in his routine, and the resulting speed of his "flying visit," as one of his sisters later called it.[1]

It wasn't until June that Melville returned to New York to work closely with the printers as he put the final touches on his book. After his winter and spring in the Berkshires even June seemed too hot in what he called "the babylonish brick-kiln of New York." So he quickly retreated to Pittsfield "to feel the grass—and end the book reclining on it, if I may."[2]

Sarah was already enjoying her new home in June, thankful that she didn't have to remain in New York for the summer. In a letter to one of her English in-laws, she echoed Melville's comments about the contrast between the city heat and the more relaxing weather in the Berkshires. "While the denizens of the city are complaining of the intense heat—we up here amid the hills are rejoicing in cool breezes and balmy days." As in the previous season, she seemed to thrive in the Berkshires. Rowland had his business in New York to keep him busy, and became so caught up in his work that he rarely visited Broadhall that summer. To the family in England, Sarah wrote, "Rowland is well but looks much thinner than when we landed—he has been very much engaged too of late. I trust when the summer has gone he will regain his fat a little."[3]

More or less on her own for the season, Sarah filled her new house with as many of her relatives and friends as she could find, issuing last-minute invitations for one hastily arranged event after another. The summer became a continuous party. Few went away unhappy. Friends praised her "rare talent . . . of putting her guests at ease." Fond of champagne, she served it generously at her parties, and Broadhall quickly earned a reputation as a place of "free-hearted hospitality."[4]

In July and August, as he finished his work on *Moby-Dick*, Melville was able to spend more time with Sarah, and gradually he lost himself in the whirlwind of her social life. When a summer shower disrupted one of her picnics, he led her little party to the shelter of a loft in a neighbor's barn. The group spread out in the hay and passed a pleasant hour talking among themselves and listening to the patter of the warm raindrops on the roof. At one point Melville entertained the group with an exuberant recitation of a friend's patriotic poem about the Revolutionary War. He was funny and charming, and not at all like the troubled man who had spent the winter writing all day in isolation. At sea, he had often enjoyed such moments of uninhibited fun, but rarely on land since his last voyage—except with Sarah.

Two days later, there was a musical evening at Broadhall, with singing and three or four performances on the piano by some of Sarah's friends. It went on until midnight and Melville stayed until the very end, riding the short distance back to Arrowhead under a bright summer moon. His wife didn't take part in any of the summer festivities this year. She was having a hard time with her pregnancy. Living with her difficult husband and his large family had never been easy, but now that he seemed determined to spend so much time with Sarah, Lizzie found herself confined to the house and unable to do much of anything. She was almost entirely dependent on her husband's mother and sisters for company. Writing to her own family in Boston one hot summer day, Lizzie found that her frayed nerves were getting the better of her. "I cannot write any more," she suddenly admitted in the letter. "It makes me terribly nervous—I dont know as you can read this I have scribbled it so, but I can't help it."[5]

She couldn't have been pleased when Melville announced in early August that he was going to make an overnight trip to the top of Mount Greylock with Sarah and a few of her friends. By now Lizzie must have had her suspicions about the amount of time her husband was spending with their pretty neighbor, but Melville could have explained away the trip as an innocent bit of fun. Melville's brother Allan was going to come along, as well as his sister Augusta. Evert Duyckinck—visiting again from New York—was bringing his younger brother George. There would be ten people in all, but neither Sarah nor Melville would be going with their spouses, and the plan was to spend the night under the stars at the summit.

In fact, this expedition to Greylock was a scandal, one that would generate so much gossip that Sarah would proudly refer to it in print as "that excursion to Greylock," as though everyone in Pittsfield knew about it. The most daring thing she ever wrote was a short chapter about it in a little book published by her friend J. E. A. Smith. The book was supposed to be a harmless guide for local tourists, but Sarah turned her contribution into a defiant celebration of a forbidden night of pleasure. Without explanation, Smith eliminated it from the next edition. Using an older spelling for the mountain range that included Greylock, Smith called his book *Taghconic: or, Letters and Legends about Our Summer Home*. It was published in 1852.

Sarah didn't seem to care much about what others thought of the excursion. She wanted to enjoy herself, and in her almost manic urge to make the most of each day, she resented having to abide by conventional morality—or, as she called it, "the iron rule which cramps and confines our best and purest feelings." The more traditional members of the community were shocked by the outing and were eager to condemn it as—at the very least—a reckless temptation. After venting in *Moby-Dick* his own frustrations with the "iron rule" of civilized society, Melville was more than ready to see this adventure as a temporary escape from a world that seemed to thwart natural desire. It was a chance to be a castaway once again, a free man at the top of a mountain. He had a word for roaming the Berkshires like a castaway. He called it "vagabondism," and earlier in the summer he had suggested that Hawthorne try it with him. "You and I," he wrote, "must hit upon some little bit of vagabondism, before Autumn comes. Graylock—we must go and vagabondize there."[6]

After staring at Greylock all winter long, Melville wanted to set foot on its summit and show that he had overcome obstacles of all kinds. He would conquer Greylock as a prelude to conquering the literary world. Again, however, Hawthorne wasn't ready to play that game, but Sarah was, and—like her effort to crown him with a laurel wreath later that year—this trip to the mountain was her tribute to Melville. She had never attempted anything like it before, and wouldn't try again.

MELVILLE'S FAMILY MAY HAVE HOPED that the trip wouldn't incite any gossip in the village, but it was inevitable that people would talk about the latest exploits of Sarah and her friends when the biggest landmark in the region was the destination of their overnight adventure. Sarah took the precaution of hiding her own identity in her essay for Smith's book, calling herself simply a "friend" of the editor. She also didn't name any of her companions, and left the composition of the group vague, saying only that the individuals were of "kindred mind, taste, and feeling." But many local people, including Melville's mother, soon learned that Mrs. Morewood was the author. In the matter of the trip itself, Sarah was unapologetic, presenting it as an exercise in romantic freedom. For her, the beauty of the area was a call to follow the dictates of nature and enjoy life with less restraint. The message of Greylock was clear to her: "Commune with your own heart, and be still." Evert Duyckinck had the impression that Sarah was managing much of Melville's time that summer. "The order of the day is a drive & call upon Mrs. Morewood," he wrote from Arrowhead in early August. "What next depends upon her kind and inventive genius."[7]

To provide at least an appearance of propriety for the overnight trip, Sarah invited one of her sisters to come along, as well as a visiting friend named Miss Henderson, and a local clergyman—George Entler. Rev. Entler was a dubious choice. The same age as Melville, he had already been dismissed from two Congregational churches— one of them on the outskirts of Pittsfield at the village of Windsor. He would eventually abandon his religious calling to teach German. He wasn't attached to any church when he accompanied Sarah's party to Greylock, and afterward he would never have another ministry in the Berkshires.

Allan and Augusta were Herman's closest siblings, but even they may not have fully realized how the trip would look to the gossips in town. Sarah regarded Augusta as "warm hearted . . . and more romantic than most people suppose her to be." As such, she was the perfect person to add respectability to the trip without dampening the pleasure of it, but she was under no illusions about Sarah's power to overwhelm with charm anyone who came into her orbit. "You are one of God's brightest creatures," she later told her, "gifted with powers of mind & a fascination that wins love & confidence, that attracts & interests."[8]

Rounding out the party was Allan's wife, Sophia, and the Duyckinck brothers, all of whom seem to have been swept into the excursion without much choice. They were guests at Arrowhead and must have felt obliged to go along. Like Augusta, George Duyckinck was so above reproach that he added a respectable air to the adventure. A devoutly religious bachelor, he was the most innocent-looking one of the group. Slightly built, with a receding hairline, a pale complexion, and an unimpressive beard that barely covered his jaw, he lived with his brother and his sister-in-law, and would never marry, immersing himself in religious biographies and eventually writing a few of his own. For now the *Literary World* was his main focus, but he was soon to become treasurer of a high-minded group not known for encouraging secret romantic encounters on mountaintops, the General Protestant Episcopal Sunday School Union and Church Book Society.

With her party assembled, Sarah threw herself into this adventure wholeheartedly, arranging for ample supplies of food and champagne to be included in the outing, and working out the best routes for the ascent. The group was divided in two, and the parties arranged to meet near the summit after approaching it from different directions, using wagons at first, and then going the rest of the way on horseback and foot. Melville was in Sarah's party. All in all, it was a climb of more than three thousand feet, and not a minor undertaking for any New England lady of the period.

It's possible to re-create their ascent, and though any reasonably fit person would find parts of the climb undemanding, there are steeper places that require more than average strength. Modern climbers may reach the top with little difficulty, but it's worth keeping in mind that Sarah and the other women in the party did it in Victorian garb and old-fashioned boots. As it was, even some of the men were exhausted by the ordeal. Evert Duyckinck was one who didn't fare well. Struggling to finish the climb, he was "gasping with open mouth and dripping at every pore" as he came up the trail. "The last mile of that is tough," he later complained. "I was a blown horse when we reached the summit."[9]

Mountain-climbing ladies were indeed a rarity in the Berkshires of the 1850s, but it was unheard-of for a woman—especially a wife and mother unaccompanied by her husband—to spend a night at the top of one. A few of the more liberal-minded citizens of Pittsfield couldn't help but admire Sarah for her spirit in undertaking such a challenge. Dr. Holmes was one of those. In his novel inspired by her, Elsie the heroine is distinguished in part for her midsummer rambles to the local mountain. "Elsie was never so much given to roaming over The Mountain as at this season; and as she had grown more absolute and uncontrollable, she was as like to take the night as the day for her rambles." From the calm safety of his summer house in the meadows below Broadhall, Dr. Holmes knew how scandalous such freedom in a woman could be, but he also couldn't help feeling that it was sexy beyond measure. Fresh from her summer "rambles," Elsie never looked so "superb." She was "never so threatening in her scowling beauty. The barred skirts she always fancied showed sharply beneath her diaphanous muslins; the diamonds often glittered on her breast as if for her own pleasure rather than to dazzle others."[10]

No one in town was quite sure what happened on "that excursion to Greylock," but Sarah didn't waste much time giving them a clue.

15

LOT'S WIFE

As mountaintops go, Greylock's is rather tame—at least in summer. The summit is broad with lots of room to wander from side to side for different views. Some of the slopes are gentle enough to encourage cautious strolls near the edge of the mountain. What mere photos can't convey is the sense at the top of standing in a snow globe at rest—a tranquil, watery atmosphere of blue and white where everything around the mountain seems to be floating in and out of the hazy distance. The wide grandeur of the green countryside below also seems in movement, changing with the light. When Henry David Thoreau spent the night here in July 1844, he felt that he was immersed in "an undulating country of clouds. . . . A country as we might see in dreams."[1]

It can be exhilarating to stand at the top in high summer, the air cool and pungent with the smell of balsam. When darkness falls, the light fades slowly as the horizon goes gray and then black. Modern climbers can stay overnight at a lodge with all the comforts of the world below, but in Melville's day the only shelter was a ramshackle observatory with rough places for sleeping that brought to mind the comfort of a hayloft. It was exactly what Sarah was hoping for. She loved the rustic feel of the structure, and she was in awe of the view, which was all the more impressive to those of her time, who were born long before anyone could admire the landscape from an aircraft window.

Far from exhausted by the ascent, Sarah reveled in it, stopping frequently to admire the views and the vegetation, and to gather wildflowers. By the time everyone reached the summit, the sun was going down and a summer mist was hanging over the rugged expanse below, adding—said Sarah—"beauty to what was already too beautiful for description." Their evening meal was taken in the observatory (Sarah called it the Tower), with "brandy cherries" served as a special treat. Melville—by far the strongest in the group—chopped wood for a large fire outside, and everyone gathered around it to warm themselves. Looking up, they watched the summer moon—"full and red"—rise "more majestically than usual." The whole party stayed up until midnight, talking and drinking champagne, with extra supplies of rum and port wine. Then they gathered in the observatory to sleep under buffalo robes left behind in the winter. A candle made from the oil of a sperm whale was placed in a champagne bottle to provide a little light, but it flickered out long before dawn.

Soon many in the group were sound asleep, but not Sarah and Melville, who stayed up all night. They were among the few who were "too merry for sleep," as she put it. Her laughter kept waking up Evert Duyckinck, who muttered wittily, "Sleep no more, Morewood has murdered sleep." Casting her gaze across the awkward forms of her slumbering friends, Sarah liked the unguarded, casual atmosphere of the night, and later wrote of it, "How absurd it is, when parties go on such wild excursions as this one was, to expect reserve, or any of the etiquette of refined life."[2]

After so much excitement and drinking Sarah and Herman didn't spend this romantic summer night at the top of a mountain making polite conversation. They did what would have come naturally to two people in love, taking advantage of the late hour and the darkness to enjoy a passionate bond that had been growing for more than a year. This was their reward for all the months they had endured apart, and for the book that Melville had created in her absence with Greylock always in view. Herman was certainly keyed up for this moment. He had been doing his best throughout the trip to show off his virility, as if that were necessary, and Sarah had taken careful note of his various feats of strength, especially when he quickly climbed a tree on the way up to signal to the other half of their group. Describing this scene later in her essay for Smith's book, she made a point of capturing the sheer joy of Melville's antics: "Suddenly we are startled by shouts which echo through the wood like the yells of the red men, and one of our party, with the agility of a well trained sailor (as he was) soon ascends the trunk of a tall tree, and from a seat which appears to us dangerously insecure, echoes shout for shout, till the remaining few of our party . . . make their appearance."[3]

All this manly exertion likely filled Sarah with amorous thoughts. After Melville built the fire at the summit, she grew excited as she watched the flames flicker across the faces of her party. The fire reminded her of those moments of reverie that Rev. John Todd had warned would unduly stimulate the mind to impure thoughts, and of the competing author ("Ik Marvel") whose *Reveries of a Bachelor* advised surrendering to temptation. "So a large fire is lighted under a giant stump," she recalled in her essay, "and we gather about it, each one indulging, like Ik Marvel, in his own reveries."[4]

At the end of her essay Sarah cleared up any doubts about her state of mind on the mountain that night when she lamented having to leave it the next day. If she had wanted to play it safe, she could have said Greylock was an Eden, and that she had hated to see a perfect adventure come to a close. Her night in Melville's company brought to mind something more sexually thrilling than an innocent Adam and Eve. What she had in mind was a biblical place so notorious that even the vaguest reference to it was sure to rattle Pittsfield to the core. The wonder is that she talked Smith into allowing her to compare her night on Greylock with a night in Sodom. Confessing that on her morning descent she kept looking back at the mountain, Sarah wrote in *Taghconic* that she didn't want to leave because what she had enjoyed on the summit was a kind of freedom forbidden in the world below, with its "iron rule" of morality interfering with "our best and purest feelings." On her way down the mountain, she wrote, she felt "like Lot's wife, casting many a lingering look behind."[5]

In a region once dominated by the strict code of Puritan demagogues, this was a stunningly rebellious note of blasphemy to outdo anything close to it in *Moby-Dick*.

Like the woman turned to a pillar of salt for looking back longingly at sinful Sodom before it was obliterated, Sarah was refusing to ignore pleasures simply because they violated laws or commandments. She was willing to boast of her defiance at a time when no woman of her standing would ever have dreamed of comparing herself to Lot's wife. As Ahab says, "Thy right worship is defiance."[6]

Obviously, the new mistress of Broadhall was as unashamedly

forward as Hawthorne was reticent and aloof. She didn't live six miles away from the most handsome man in the neighborhood—she was six minutes away on horseback. Sarah was emboldened by Herman, and her sympathy and understanding inspired him in turn to share her defiance, attacking religion, authority, and civilization itself with far more force and spirit than ever before. It's partly why he felt that he could make a proud boast of his own to the author of *The Scarlet Letter*, telling him that the forthcoming *Moby-Dick* was not only a "wicked" book, but one with a secret motto celebrating baptism in the name of the devil ("in nomine diaboli"). Hawthorne must have thought such talk was mere bluster or a sailor's rant, but Sarah knew intimately the depth of Melville's dissatisfaction with the well-ordered, self-satisfied world that would condemn him for loving her, and for wanting to write books undermining its values.[7]

In his anger and frustration, Melville wanted both to succeed and to spurn success. He hated debt, but couldn't escape it. He lived like a ghost in his own cramped home all winter, but burst into life the moment Sarah returned. His wife was pregnant, but he would rather spend the night on a mountain than in bed with Lizzie. Though he needed her father's money to live, he didn't like admitting it, and increasingly resented it. All these contradictory urges and circumstances were becoming harder to manage except when he could escape them with Sarah. In the weeks after "that excursion to Greylock," the couple turned their August night on the mountain into a glowing example of the freedom they craved, the ultimate escape from family burdens and career disappointments. The experience became, said Sarah, a "shrine for memory to return and refresh itself at, when cares and trials make us weary." To Evert's brother, George Duyckinck, Sarah reported that she and Melville were still reliving the experience on the mountain: "Greylock is not forgotten here but often recalled in an amusing way—by Mr Herman or myself—In some of our long walks we have taken a spyglass with us so as to bring nearer to us the Tower and its associations."[8]

WITH THAT SPYGLASS tucked under his arm, Melville could indeed think that he had arrived in the Berkshires by sea and was a castaway again, and with Sarah at his side to explore their paradise, rather than Fayaway. Confiding such intimate information to the incorruptible George was part of the fun of being "wicked" together. Just as Sarah liked to tease upright men like Dr. Holmes with her beauty and charm, and to befriend wayward clergymen like Rev. Entler, so she would in the months after Greylock try to captivate poor George.

It was easy to shock him, and she loved doing it. He was almost too easy to fool. A friend once described him as "meek" and "guileless," and "detesting wrong & deceit." One day Sarah made a passionate plea to George to stop shaking her hand limply when they met and to offer "a warmer grasp." Knowing that even such mild teasing would upset him, she wrote gleefully, "[I] can almost see how shocked you are looking while reading this letter."[9]

At first George seemed to misunderstand the nature of her attentions, responding by sending her religious books for study and stiffly offering them to her in a tortured rhetorical question: "Will Mrs Morewood please accept the accompanying little volumes . . . as a slight memento of Greylock and mark of heartfelt sympathy respect and gratitude [?]" In time he realized that she was trouble, and he did his best to avoid her. "You refused to call on me on Monday," she wrote indignantly after he kept finding excuses not to see her in New

York. Another letter began, "I cannot at all understand the reasons why you treat my letters with Silence and I am deeply pained that you do so."[10]

Dr. Holmes had a theory for why Sarah was so provocative. In the sexist thinking of the time, he concluded that his fictional Elsie Venner was dangerously alluring simply because she couldn't be otherwise. It was in her nature, as if she had been poisoned at birth by some slow-acting venom. Holmes wanted to hate the sin and love the sinner, turning his heroine into "a proper object of divine pity, and not of divine wrath." A schoolmistress in his novel says of Elsie, "Women's love is fierce enough, if it once gets the mastery of them, always; but this poor girl does not know what to do with a passion." No doubt Sarah would have argued that she was capable of managing her passions, and she tried to prove it in loving Melville. When a sympathetic but innocent male character asks Elsie what he can do to help her, she has a simple answer: "Love me." Yet in her small town—with all the complications and entanglements of marriage and courtship—it is too much to ask of any man already spoken for.[11]

Toying with respectable George may have been a game that both Sarah and Melville enjoyed. George's oldest friend was William Allen Butler, the man whose bride Herman "kidnapped" the previous summer. Just as Butler was led on a wild chase, so George would find himself pursued to laughable extremes by Sarah. Valiantly he resisted, and not only with her, but with every woman who entered his life. He wouldn't accept Sarah's invitations to the Berkshires and played hide-and-seek from her in New York. He wasn't experienced enough with women to know whether Sarah was playing with him or setting a more serious romantic trap. Everyone knew that his only serious interest was his faith. As a friend once said of George, "Everything about him proclaimed his religion to be *life* . . . devotion forming an essential part of his disposition."[12]

To deflect attention from her relationship with Melville, Sarah would soon find it convenient—and probably amusing—to pretend with Herman's sisters that banal George was the only man for her. The credulous sisters and their sanctimonious mother swallowed the bait entirely, and began worrying that she would somehow compromise her honor with young Duyckinck. If nothing else, this ruse shows how completely the family misjudged their own Herman, believing that the spirited Mrs. Morewood saw nothing in him, but everything in a pallid character like George.

What she really wanted all along from the young man—who was only a month her junior—was his help in getting her poems published in New York. She expected him to take her seriously as a poet and to keep her informed of the latest book news and literary events. He rudely dismissed her early efforts to send him anything. In a letter beginning "My dear Sir," she responded politely, "I do not send you the verses I wrote you about because you told me not to trouble myself in so doing. It was well you told me in time—else I might have inflicted upon you more reading than would have been agreeable to you."[13]

It was when George responded so coldly to her verses that she began writing to him more warmly, hoping to win over his heart if not his head. Cleverly, she filled her letters with lush descriptions of the natural world to show off her poetic talents even when he wouldn't look at her actual poetry. When she read Evert's remark in the *Literary World* that "we hold every production of the mind to be of interest, like a collection of minerals," she promptly quoted it back to George and suggested that her letters were worth collecting, too.[14]

Nothing could overcome his firm resolve to be good, no matter how many suggestive poems Sarah sent his way. Her letters to him fill a small folder at the New York Public Library, and the tone is at times so literary that they almost sound like rough drafts of a novel. At a time in New York when an enraged husband might choose to thrash his wife's lover in the street, George must have been stricken with fear when Sarah suddenly sent him a fervent invitation to meet her husband. She made it sound as though they would have an affair, and they would have to inform Rowland. "I shall find very great pleasure in bringing him [Mr. Morewood] to see you. I want you to know him and to understand that I act with his full knowledge and wish and will—a fact which those *who know me* ought to know." Though an affair was the last thing George wanted, Sarah's letters had to worry him for the damage she might do to his reputation. It would only make sense if he blamed Melville for her unwanted attentions. Herman introduced her to George, and he seemed to be her closest companion in the Berkshires, the one with whom she shared vistas through that spyglass.[15]

Whatever the Duyckinck brothers thought of their old friend Herman after "that excursion to Greylock," it couldn't have been good. In November, a change in their opinion of Melville seems to have triggered an otherwise inexplicable attack on him in the pages of their influential journal. It would have devastating consequences.

16

ALL FOR LOVE

Throughout that emotionally charged summer and into the early autumn of 1851, Melville couldn't see enough of Sarah. There was card playing and dancing at Broadhall, and harvest festivities in the neighborhood. On several occasions, they visited one of the larger lakes in the area, Pontoosuc, where they went boating and fishing. The shoreline was bordered on one side by low hills that stretched toward Greylock, and the blue water sparkled in the sunlight. Melville confided that it was fast becoming one of his favorite places in the Berkshires, telling Sarah "that each time he came there he found the place possessing new charms for him."[1]

About a mile from the lake was one of her favorite spots, a pretty grove where a huge limestone boulder sat on a much smaller base of rock, balanced there by nature ages ago but looking as though it might tip over at any moment. Sarah loved it for its dramatic effect. Her guests could always be amused on a trip there when she seemed to defy danger by lying under the overhanging edge of Balance Rock, as it is still called, or by pretending that she could push it over. A friend once described her as rushing up to the boulder "with a merry laugh, declaring she would push the monster from the seat he had kept longer than was right."[2]

Visiting it with Melville and others, she enchanted him by hiding a music box under the rock and acting as if the tune magically flowed from the earth itself. This was the kind of fanciful act that stirred his imagination and set him to thinking of similar wonders in his reading. What came to mind was the story of the Egyptian statue that made a low moaning sound at dawn—a noise almost musical—and was supposedly a monument to an ancient king, Memnon. Listening to Sarah's music box while their friends picnicked, Melville decided to christen the rock Memnon, and to carve that strange word into a nearby tree filled with the names of couples who had been there before them. In this way the name united them and was a clever substitute for inscribing "Herman" and "Sarah" alongside those other lovers.

Caught up in so many vivid scenes from day to day, Melville couldn't help but think of writing about them. This was the greatest story of his life, unfolding before his eyes with both drama and suspense. He could only guess where this story was headed, and he knew that he couldn't afford to tell everything. Because he'd always drawn from his experiences to create his books, here was a sequence of events in a beautiful location that was a narrative gift impossible to ignore. The trick was to hide identities without ruining the drama.

With *Moby-Dick* completed in August, and now awaiting publication in November, Melville could have waited months to start a new book. He wasn't sure how his whaling saga would be received, and writing books was his only way of making money. Just as Sarah couldn't wait to write about Greylock, so that memorable night was probably the spark that started Melville writing the Berkshire romance that would become his seventh book in six years—*Pierre; or, The Ambiguities*.

HE WROTE *PIERRE* with white-hot speed not only because he was in his usual hurry to sell a new book, but because he was chasing the events of his story as he lived them, trying to put them on paper even as fresh dramas hurtled toward him. The basics of his novel were taken straight from his life, but with a major twist that was so unusual and controversial that it threw readers off the track of the real story for more than 150 years.

Without this strange twist, it's a fairly straightforward tale of an idealistic youth whose life is forever changed by his romance with a dark, mysterious beauty. Isabel is an unconventional young woman who lures the budding author Pierre away from his steady girl, the good-hearted Lucy, the uncomplicated daughter of "an early and most cherished friend of Pierre's father." In other words, Lucy is essentially Judge Shaw's daughter, and Isabel is Sarah. The parallels aren't exact—they never are in good fiction—but the basic arrangement is there, and so are many of the actual elements of character and setting. For most of the book Pierre is torn between these two women, both of whom are devoted to him, and neither of whom he wants to hurt. The recognizable landscape of the Berkshires is featured in the story, with Greylock, the Balance Rock (called the Memnon Stone), and Broadhall and Pittsfield thinly disguised.

Then comes the twist—Isabel is not simply the attractive woman who seduces the starry-eyed hero, but someone who calls herself his secret half sister, the supposedly illegitimate daughter of an affair his father—now dead—hid from the family for years. By this alteration of his situation in real life Melville fooled almost everyone, changing one kind of forbidden love—adultery, in his case—to the more sensational and sinister suggestion of incest in the novel. In doing so he added a sexual layer that scandalized his readers beyond anything he anticipated. They could confront adultery, but incest was the great unmentionable.

It was also a wonderful red herring. Scholars have scattered in all directions trying to establish whether Melville had a secret sister or even an especially amorous cousin, and every lead has hit a dead end. Undeterred, some have suggested that Melville's mother must have been his forbidden lover or that Isabel is actually a "he"—a Nathaniel Hawthorne in petticoats, disguised in order to hide Melville's secret love for him. There is no question that Herman yearned for a closer relationship with the older writer, but there is also no evidence that he contemplated having a physical relationship with him. For one thing, Hawthorne couldn't bear the idea of such a relationship. Advocates of a supposed love affair between the two writers seem not to have read Hawthorne's tirade in 1851 against men he saw living in close intimacy at a Shaker dormitory that he visited with Melville in the Berkshires. The mere idea of men sharing such close quarters so filled him with horror that he not only recoiled from it but wanted the "disgusting" Shaker men to disappear from the face of the earth. In his journal Hawthorne lashed out at the group: "Their utter and systematic lack of privacy; their close junction of man with man, and supervision of one man over another—it is hateful and disgusting to think of; and the sooner the sect is extinct the better—a consummation which, I am happy to hear, is thought to be not a great many years distant."[3]

Once Sarah's place is established in Melville's story, *Pierre* isn't much of a mystery. Even filtered through the murky lens of its incest theme, Isabel's spellbinding effect on Pierre is the same as Sarah's hold on Melville. The story is like an eruption, a great outpouring of emotion after long delay. At last Melville found a way to explain the transformation in his life that had abruptly caused him to settle in the Berkshires, sent him deeply into debt, strained his marriage, and inspired in *Moby-Dick* the best book he would ever write.

REDUCED TO THE ESSENTIALS of its love triangle and its hero's literary struggles, *Pierre* brings to life Melville's mood of desperation and exhilaration in this period. Like him, his hero is haunted by the "beautiful sad-eyed girl" with her "long, dark locks of mournful hair." The combination of her "dark, regal" bearing and mysterious sensuality have an electric effect. Rising to poetic heights equal to those of *Moby-Dick*, Melville describes his heroine as a kind of storm rising on the horizon and rapidly engulfing his hero in a burst of overwhelming energy. "She seemed molded from fire and air, and vivified at some Voltaic pile of August thunder-clouds heaped against the sunset."[4]

There was no one like Isabel in the author's life except Sarah. What Melville says of Pierre's literary labors was also true of his: "He seems to have directly plagiarized from his own experiences to fill out the mood of his apparent author-hero." As Pierre's emotional life becomes increasingly complicated—with Lucy the "good angel" on the side of right and society, and Isabel the supposedly "bad angel" on the side of love and desire—the ambitious young writer vows "to give the world a book, which the world should hail with surprise and delight."[5]

Three overwhelming problems confront Isabel and Pierre. They can surrender to love, but sex is taboo; they have a secret they relish, but can't afford to reveal; and they yearn to escape the morality that has so restricted their freedom, but they can't. Standing in their way is what Sarah had called the "iron rule" that interferes with "our best and purest feelings." They are trapped whichever way they turn. Secrecy becomes a weapon they can employ to confuse the moralists determined to condemn them. "The deceiving of others" must be done, they agree, "for their and our united good." Creating an invisible fortress of secrecy around them, Pierre becomes her faithful knight, vowing to "champion Isabel, through all conceivable contingencies of Time and Chance." Behind this fortress, they allow themselves slowly to test the limits of what is wrong and right.[6]

The easy way out for Pierre is to walk away from what is forbidden and accept the safe love of Lucy, the well-bred woman from a respectable family. Though Lucy has her appeal, Isabel shimmers in his eyes with "sparkling electricity" and touches the core of his being. The forbidden nature of their attraction makes it defiant in a way that Pierre finds endlessly thrilling. When Isabel vows not to abide by any "terms from the common world," there is almost a kind of swoon in the narrative. There aren't any whales in sight or lashing gales on an open sea, but Melville writes of this moment as though describing the emergence of some mythic creature from the waves. In her proud defiance even Isabel's hair is "scornful"— "Her changed attitude of beautiful audacity; her long scornful hair, that trailed out a disheveled banner; her wonderful transfigured eyes, in which some meteors seemed playing up; all this now seemed to Pierre the work of an invisible enchanter. Transformed she stood before him; and Pierre, bowing low over to her, owned that irrespective, darting majesty of humanity, which can be majestical and menacing in woman as in man." As Dr. Henry Murray noted long ago, "Isabel may be roughly correlated with the 'ungraspable phantom of life' reflected in the sea; and Lucy with the securities, comforts, and consolations of land."[7]

For a man who had little experience writing love scenes, Melville rose to the challenge when his lovers cross the line between affectionate warmth and real passion. Leaning into each other, they begin to press so close with their lips and hands that they become almost one body. "They changed," Melville writes with great feeling and simplicity; "they coiled together." He writes about lovemaking as though he is discovering its pleasures for the first time. In his new enthusiasm he declares that "love is god of all. . . . This world's great redeemer and reformer." In case anyone doubted the source of all his excitement, Melville traces it to a renewed appreciation of women as love's "emissaries," a grateful sense of wonder for their refining sensibilities in a world sorely in need of them. As one who had begun his writing career gazing admiringly at the female form in a South Seas Eden, Melville opens his eyes now and finds that the woman makes the paradise, not the paradise the woman: "Where a beautiful woman is, there is all Asia and her Bazars." Whatever else this is, it is the sound of a man in love. It isn't heard in Melville before *Moby-Dick* and *Pierre*, and it won't be heard in his prose after those works. By comparison, Fayaway was merely an infatuation.[8]

For anyone who has loved and was led to believe it was wrong, Pierre asks the great question of the novel, "Is Love a harm?" The answer for society is always complicated, depending on the nature of the love, but from deep within the actual moment of a romance as intense as Melville's, there is only one answer—a resounding no. Society won't abide that answer in so many cases, and that is the tragedy for those who choose to love anyway.[9]

PART OF PIERRE'S ATTRACTION TO ISABEL is that she so closely resembles the pale beauties in romantic fantasies of tragic love, the woman with raven hair and "bewitchingly mournful eyes." In their determination to love each other despite the condemnation of the world they are as doomed as Ahab sailing into his vortex. It is as if a malignancy is ever present and always posed to strike. Isabel seems to think that death is close at her heels, and that his shadow can be detected in her appearance. "Look," she tells Pierre, "see these eyes,—this hair—nay, this cheek;—all dark, dark, dark.... Was ever hearse so plumed?"[10]

The world expected Melville to be the manly seafarer with endless tales of ships, storms, and island adventures, and so there were few readers prepared to take this very different voyage into Melville's turbulent emotional life on dry land. But it was a book that he had to write, though he was increasingly doubtful it would succeed. It wasn't just that readers would be confused to find their stalwart mariner suffering from lovesickness in an unfamiliar land far from any sea. It was also because the author knew he couldn't tell the whole story. His characters would only be ghostly players reenacting a lesser version of the real drama that had been unfolding in the Berkshires for the past year.

17

HARVEST

On his way to attend yet another party at Broadhall, Melville was stopped by someone from the village and handed a letter. It was mid-November 1851, *Moby-Dick* had just been published by Harper & Brothers in New York, and reviews were starting to appear in the press. The envelope didn't contain news of any professional critic's latest pronouncement. It was a letter from Nathaniel Hawthorne giving his private views of the new book. Nervous with anticipation, Melville raced ahead to Broadhall and read it there. He couldn't wait to hear the verdict.

The letter was good. Hawthorne liked the novel and was liberal with his praise. Melville was euphoric, believing that if he received such a major writer's approval for his masterpiece, everything else was destined to fall into place. Unable to restrain his joy, he sat down the next morning to write a letter of gratitude. After his long struggle, he felt vindicated and—at least for now—at peace with himself. "A sense of unspeakable security is in me this moment," he wrote to Hawthorne, "on account of your having understood the book. I have written a wicked book, and feel spotless as the lamb. . . . I would sit down and dine with you and all the gods in old Rome's Pantheon." If Hawthorne had been embarrassed before by some of Melville's warm declarations of literary friendship, this letter was sure to make him blush. His friend was ready to turn him into a lord of the universe, and join him on the cosmic pedestal. "I feel that the Godhead is broken up like the bread at the Supper," he told him, "and that we are the pieces. Hence this infinite fraternity of feeling."[1]

At this moment the world was too small for Herman Melville. He knew that he had created a masterpiece, a book of unparalleled boldness and brilliance with soaring passages of prose like the best poetry. In a little village far from the great capitals and commercial centers of the world, he had fashioned a work of genius in the solitary space of his farmhouse study. His family needed him and wanted to believe in him, but they had no concept of the vast range of his imagination and the breadth of his learning. For the most part, to his family and to almost everyone in Pittsfield, he was just Herman, the eccentric and willful literary man who had settled on his Berkshire farm for no logical reason, and who was staking his fortunes on a strange book about whales.

Yet it was his ambition to shake the world—to reinvent the novel, remake American literature, reintroduce the world to the wonders of the whale, and to redesign J. M. W. Turner's artistic style to shine anew in prose. Few writers have come as close as Melville to achieving such high goals in such modest circumstances. Like most geniuses, his chief assets were his talent and his yearning to create something never seen before, but working in an environment where so few of those in his daily life shared his high sense of purpose was a constant weight on his spirits.

All of which explains why Hawthorne and—to a greater extent—Sarah Morewood were so important to him, and why he wanted each to glow in his imagination like a companion star. They helped to make the isolation bearable and to make him think that impossible goals were worth chasing. In his greatest moments of enthusiasm, he could elevate them to a lofty plane where they could stand in majesty apart from the herd. Just as Sarah could be celebrated as the "ever-excellent & beautiful Lady of Paradise," so Hawthorne's appreciation for the larger designs of *Moby-Dick* earned him a celestial ranking. As Melville put it, "You were archangel enough to despise the imperfect body [of the book], and embrace the soul."[2]

It's only because we have a surviving transcript of Melville's letter that we know what was in Hawthorne's. Unfortunately, like so much of Melville's correspondence, this landmark letter of praise has disappeared, probably because of what Melville once called "a vile habit of mine to destroy all my letters," so it is impossible to know just how encouraging Hawthorne was or whether he hedged his approval in any way. As a writer more attuned to the marketplace than Melville, the successful author of *The Scarlet Letter* and *The House of the Seven Gables* must have known that *Moby-Dick* was going to be difficult to sell in America. Yet he also knew that Melville didn't want the usual warnings about the tastes of the reading public, and the fickle attitudes of critics and publishers. What he wanted was apparently exactly what Hawthorne gave him—one literary giant's blessing to another. It wasn't Hawthorne's kind of book, but it was enough—as Melville acknowledged—that he "understood the book." More than sales and more than simple praise, the author wanted to know that he had succeeded in doing what he had set out to do. To Hawthorne's great credit, he gave him that assurance.[3]

Sending his letter of praise was also a graceful way of exiting the Berkshires. As another winter approached, Hawthorne and his family were ready to clear out. They were headed to the other end of the commonwealth and would never reside in the Berkshires again. Writing to Melville was almost the last thing Hawthorne did as he closed up his rented cottage in Lenox. If he had waited much longer, he might have completely undermined Melville's idealized image of him by angrily revealing what he truly thought of this rural paradise. It would have broken the younger writer's heart to know how his beloved Berkshires rated in Hawthorne's estimation. "I detest it! I detest it! I detest it!!!" stormed Hawthorne in his diary. "I hate Berkshire with my whole soul, and would joyfully see its mountains laid flat."[4]

Given the haste of his exit, he didn't have much time to write a proper review for publication, but Melville was quick to let him off the hook, telling him that he didn't need to take the trouble. If *Moby-Dick* had passed muster with Nathaniel Hawthorne, then surely the best critics would find much to praise on their own, or so Melville must have reasoned as he waited for reviews to appear. Based on past experience, and taking account of his own fears, he was prepared to be disappointed, but Hawthorne's praise gave him hope that this time the world would correctly judge his worth. (At any rate, it would have been awkward for the older author to publish a review of *Moby-Dick* because of Melville's generous dedication page: "In token of my admiration for his genius, this book is inscribed to Nathaniel Hawthorne.")

AT FIRST, the publication of *Moby-Dick* was a mixture of good and bad news. Some of the most influential papers offered praise. The *New-York Tribune* said it was his best book; *Graham's Magazine* called it a work "of a bright and teeming fancy"; the *Philadelphia American Saturday Courier* hailed it as "decidedly the richest book out"; and the *Spirit of the Times* in New York said it was a work "of exceeding power, beauty, and genius."[5]

Few of the major American reviews took a neutral position. Most critics either loved the book or hated it, and those who hated it were unsparing in their condemnations. The *Southern Quarterly Review* damned it as "sad stuff, dull and dreary, or ridiculous." It was "not worth the money asked for it," said the *Boston Post*. Ahab was "a perfect failure," claimed the *Albion* in New York, "and the work itself inartistic." The *New York Independent* confessed that the waste of so much talent on such profane material made "us ashamed of him that he does not write something better."[6]

Melville could have used someone as influential as Hawthorne to shift the balance of opinion in his favor. As it was, he relied on Evert Duyckinck to champion *Moby-Dick* in the pages of the *Literary World*. It was only natural to assume that his old friend would be an even more generous and sympathetic reader than Hawthorne. For months Melville had been telling Evert about the book, and twice he'd entertained him in the Berkshires, providing rooms and food, and a wagon for travel.

But something went wrong. In the months since the Greylock trip, Evert and George Duyckinck had found reason to think that their gracious host in the Berkshires needed chastising. On November 22 the *Literary World* aimed its guns at the imposing bulk of *Moby-Dick*. Though Evert seems to have written the review, George was probably involved in the editing, for the heavy moral criticism suggests his influence. Though they sprinkled praise throughout the review, the Duyckincks went out of their way to cheapen the work by comparing it to an overlong "German melodrama," with "Captain Ahab for the Faust of the quarter-deck." In a snide and condescending tone, Evert treated Ahab as more of a bore than an inspired creation. He told his readers that he was so weary of spending time in the book's "melancholic company" that he now understood "why blubber is popularly synonymous with tears."

He tried to make these criticisms seem playful, but of all people, he and George should have understood how seriously Melville took his work, and how much he had sacrificed for it. It was one thing to point out weaknesses in the book. What was painful for Melville was to see friends make light of the high ambitions of the work, undermining its tragic majesty by treating it as little more than tedious melodrama. A minor writer like Evert probably felt that Mr. Melville of little Pittsfield was trying too hard to be a major writer and needed to be brought down a notch or two.

He and his brother may also have decided that Sarah Morewood's friend was becoming a little too irreverent and unconventional for his own good. The Duyckincks saved their strongest criticism for an attack on the morals of Melville's book. Evert objected to its "piratical running down of creeds and opinions," and declared with self-righteous pomposity, "We do not like to see . . . the most sacred associations of life violated and defaced." It is tempting to read this moral outrage as a barb directed at Melville personally, and not just as a criticism of Ishmael's fondness for his pagan friend Queequeg or other impious moments in Ahab's defiant assault on the universe. Religious critics could find reasons for saying that *Moby-Dick* "defaced" "sacred associations," as the New York *Churchman* did when it attacked the book for its "sneers at revealed religion and the burlesquing of sacred passages of Holy Writ." Evert's charge of violating something sacred seems unwarranted unless it was meant for the author himself.[7]

His friend was in a position to hurt Melville, and he succeeded. Even Hawthorne couldn't understand the hostile tone. He had seen Melville and the Duyckinck brothers in August, before the Greylock excursion, and thought they were all the best of friends. Bewildered, Hawthorne wrote to Evert after the *Moby-Dick* review appeared and observed politely that it "hardly . . . did justice to [the book's] best points." That was an enormous understatement. As he well knew, it was a knife in the back of a great writer who had expected more understanding from Evert.[8]

The *Literary World*'s callous treatment of a formerly valued contributor doesn't make much sense unless Evert believed—in his high-minded fashion—that he was saving Melville from himself, just as brother George must have believed he was trying to help Sarah by sending her religious books. In confidence, after the death of Sarah's precious colt, Herman had more or less admitted to Evert that he was in love with another man's wife. On Greylock that night in August one of the brothers may have been alarmed by some sign of intimacy between the two lovers, or by the later news from Sarah that she and Melville were roaming the Berkshires together with a spyglass—a "piratical" image indeed. Or they may simply have been annoyed enough at Sarah's interest in George to land a blow of their own for the moral code they valued even more than friendship.

The Duyckincks weren't malicious. They were just sanctimonious. Their review was a message from two guardians of culture and morality that Melville had done something more serious than violate good taste in an overlong melodrama about blubber. To them the sin was to undermine "the most sacred associations of life." There was nothing worse they could say of him. It's doubtful that the sins of the book were enough in themselves to merit such a devastating rebuke. The brothers must have also realized that the young man who had charmed them with the seeming innocence of *Typee* was now acting like a heathen himself, as they would have seen it, breaking what were for them the sacred bonds of marriage and family. Part of the mission of the *Literary World* was to reinforce such moral values. At least, that was the view of George's religious friends. The Episcopalian Church Book Society believed that the Duyckincks gave their journal "a savor of Church life which made it especially acceptable to the members of our Communion—opening its columns to, and inviting contributions from the younger of the clergy."[9]

For Melville, this whole episode cast a cloud over his book, and it denied him the chance to have his masterpiece taken seriously in a New York journal that so many of his fellow writers respected. It gnawed at him, as only a really bad and personal criticism can sometimes do. There was no way to undo the damage to *Moby-Dick*, and soon Melville's elation turned to bitterness.

DESPITE THE MANY GOOD REVIEWS—both in America and Britain (where the *Illustrated London News* praised Melville's "almost unparalleled power over the capabilities of the language")—*Moby-Dick* simply couldn't attract enough readers. Its sales were modest not only in America, but also in England, even though the reception there was so much more enthusiastic. In London, the publisher

Richard Bentley lost more than a hundred pounds on his edition of *The Whale,* as the book was called in Britain, using the original title

Frustrated, Bentley sent Melville a stern letter of rebuke, telling him, "If you had . . . restrained your imagination somewhat, and had written in a style to be understood by the great mass of readers—nay if you had not sometimes offended the feelings of many sensitive readers you would have succeeded in England." What was worse, he didn't see much hope for the author in the British market and didn't mince words. "Perhaps somebody ignorant of the absolute failure of your former works might be tempted to make a trifling advance on the chance of success; but . . . any new book would have an uphill fight of it." These words were like a death sentence to a young author. Of course, as the long arc of time would prove, Bentley was wrong, and Melville was right. "There are goodly harvests which ripen late, especially when the grain is remarkably strong," he had written to Bentley in 1849. He had fought his battle valiantly, and should have won it. In the aftermath of the battle, his accomplishments quickly began to look like a monumental defeat. *Moby-Dick* was threatening to sink his career, and ruin his finances.[10]

As one of his harsher reviewers declared in January, the size and scope of the new book signified nothing but the inflated self-importance of an author whose only good work was in the past: "Mr. Melville has survived his reputation. If he had been contented with writing one or two books, he might have been famous, but his vanity has destroyed all his chances of immortality, or even of a good name with his own generation." His grand calls for a new American literature with its own native giants superior even to Shakespeare now sounded hollow. For the sin of daring to write an American epic, he was damned for revealing his "morbid self-esteem, coupled with a most unbounded love of notoriety."[11]

As it turned out, perhaps predictably, much of American literary society was still too provincial and small-minded to acknowledge what Melville had accomplished. All his brilliance seemed as if it had made his life only worse. He had tried to reach the stars, and now he seemed in free fall. As he would write of his hero in *Pierre,* "He seemed gifted with loftiness, merely that it might be dragged down to the mud."[12]

AS HIS SECOND WINTER in the Berkshires began to close in around him, Melville grew increasingly desperate and angry. More than ever, he needed money. In addition to all his debts, he now had a new mouth to feed. In October Lizzie had given birth to a son. Melville decided to name him Stanwix. It was a tribute to his mother's father, an old war hero long dead, General Gansevoort, whose great victory against the British in 1777 at Fort Stanwix in upstate New York was such a revered memory in the family. It was a remarkable choice for a boy's name—awkward but heartfelt—but it seems to have been influenced by the hope that fresh glory would soon come to the family. At the time of the birth, the glow of promise still attended *Moby-Dick,* which had yet to be published in America. Anticipating success for the new novel, Melville may have thought the name was a good omen—a sign that his own literary battle would turn out well.

Instead things would take a rapid turn for the worse. The commercial collapse of *Moby-Dick* created a vortex for Melville like the one that swallowed his whaling ship. This one was about to swallow his career and leave him floating at the edge like Ishmael, his head barely above water.

18

SONS AND LOVERS

With a little knowledge of Melville's life, a detailed command of his work, and the intuition of a great novelist, D. H. Lawrence reached a few conclusions about the author of *Moby-Dick* when the book was enjoying its revival in the 1920s. In his own unique stream-of-consciousness style of criticism, Lawrence offered this general overview:

> A mother: a gorgon. A home: a torture box. A wife: a thing with clay feet. Life: a sort of disgrace. Fame: another disgrace, being patronized by common snobs who just know how to read.
> The whole shameful business just making a man writhe.
> Melville writhed for eighty years.
> In his soul he was proud and savage. But in his mind and will he wanted the perfect fulfillment of love.

Lawrence wasn't far off. Home became a kind of prison for Melville. His relationships with his wife and mother were swirling with tensions, and when they all occupied the same house, it must have been torture to balance their needs against his.[1]

When he wrote *Pierre,* the domestic arrangements at Arrowhead were in such upheaval that no one bothered to check Stanwix's birth record and correct the stupendous error of having Herman and Maria G. Melville listed as the boy's parents. No one can now explain how the name of Herman's own mother managed to be substituted for Lizzie's when the birth was recorded by the local authorities in January 1852. It wasn't a simple mistake. Someone correctly entered Maria's information—including her place of birth—despite the fact that she didn't even belong in the record. The only logical conclusion is that Herman or another family member made a curious error or that a clerk was given the impression that the leading woman of Arrowhead was Maria G. Melville.[2]

She was certainly the woman in charge when the birth was recorded, for her daughter-in-law spent all of December 1851 and January 1852 away from Pittsfield. With her newborn in her arms, Lizzie left after Thanksgiving for Beacon Hill, seeking medical treatment for a breast infection. After she was better, she showed no sign of wanting to hurry home, and Melville didn't seem anxious to have her back or to pay her even a brief visit. He was in a dark mood, and it was getting darker by the day as he worked on *Pierre* in a Berkshire winter as snowy and as bitterly cold as the last. With Christmas approaching, he was back to his solitary ways, retreating to his desk to try once again to write a book that would save his career. Now he was more alone than ever, with Hawthorne far away and Evert Duyckinck unwelcome. By Christmas Eve there was no one at home except his mother, a couple of his sisters, and his son Malcolm.

D. H. Lawrence was right when he said that Melville longed for "the perfect fulfillment of love," but Lawrence assumed that no American woman in the author's circle would have been able to cope with such a "proud and savage" man. "A mountain lion doesn't mate with a Persian cat," joked Lawrence; "and when a grizzly bear roars after a mate, it is a she-grizzly he roars after—not after a silky sheep." But, then, Lawrence never knew of Sarah Morewood, a woman who could climb a mountain and ride a colt named Black Quake, and would never purr tamely like a Persian cat. When so many people seemed to be retreating from the "proud and savage" author of *Moby-Dick*—publishers, literary friends, many of the "serious" readers of Pittsfield, and not a few critics—Sarah made a mad dash in a snowstorm from New York to Pittsfield so that she could be with Melville on Christmas. It wasn't easy for her to do this. Any Christmas dinner at Broadhall would mean inviting not only Herman, but also the rest of his Arrowhead household, and she could barely tolerate the bossy Maria Melville, who made no secret of her disapproval of nearly everything Sarah did. There was nothing Sarah hated more than being judged, yet Maria loved nothing better than judging those around her and finding fault. ("A more ungallant man it would be difficult to find," she had complained of Herman earlier that year when he rushed back home to his work rather than spending an hour waiting with her for a train at the Pittsfield station. She accused him of "dumping me & my trunks out so unceremoniously at the Depot.")[3]

In the days leading up to Christmas, Sarah was amusing herself in New York, shopping, going to lectures, and pestering George Duyckinck. For many weeks after the *Literary World* published Evert's damning remarks about *Moby-Dick*, Melville fumed over them and silently considered his future with the Duyckincks as he waited for more reviews to come in. To anyone but Melville, the negative aspects of Evert's review would have seemed less damaging when set against its carefully selected praise. As far as Sarah knew in December, George was still an innocent lamb worth poking and prodding for attention, but he was proving so elusive that she waited until the day before Christmas to abandon the chase, and to face Melville's mother in Pittsfield.

Without Sarah, the author would have spent the first Christmas after *Moby-Dick*'s publication in "a torture box," as Lawrence put it, eating dinner with his mother and sisters. He was so desperate to see Sarah that when news came of her arrival in Pittsfield on Christmas Eve, he wanted to race up to Broadhall immediately. Despite the raging snowstorm and the darkness, he brought the family sleigh to the door of Arrowhead and waited for the others to accompany him. His mother refused to budge, complaining that she wasn't going to jump up at the last minute just to accommodate Sarah's whimsical schedule. She insisted on waiting for a proper invitation to come to Broadhall for dinner the next day.

It says everything worth knowing about the pathos of Melville's life at this moment that he couldn't go see his lover because his mother wouldn't let him. She didn't have a clue that her son was in love with a married woman. It would have killed her to know that. So how did she explain why he was so eager to brave a snowstorm just to drink a toast at Broadhall on Christmas Eve? To her daughter Augusta, who was in New York, she dismissed his eagerness with the blithe explanation that Herman "loves to go out in such wild weather." As one of Melville's characters in his long poem *Clarel* would later poignantly complain, "My kin . . . would have me act some routine part. . . . This world clean fails me; still I yearn."

The scene couldn't have been more dramatic when, on the next day, Melville walked into Broadhall, stepped ahead of Rowland, took Sarah by the arm, and led her into the dining room where the laurel wreath was waiting on the plate. Actions usually speak louder than words, and Sarah's effort to crown her lover that day should have struck everyone in that room like a thunderbolt. Instead, Sarah's well-placed hints that her heart belonged to George were enough to bamboozle Maria. "What a strange woman she is," Maria later wrote of Sarah, again in a letter to Augusta. "I rather think Mister George would have felt jealous could he have seen the devoted attention paid to the author by our hostess, as he led her into the dining room, she stopt before a plate on which lay a beautiful Laurel wreath, which she gently lifted & quickly placed upon his brow."[4]

Maria knew George Duyckinck only slightly and had no idea that he would be far from jealous. Indeed, had he been present, he might have been the first to see the sacrilege in these actions at Christmas and denounce them. Sarah was pleased that difficult Maria hadn't raised any objections, and she was trying her best to get along with her. "I am strangely and strongly attracted to her and her family now that I know them so well as I do," she told George. But Maria thought that Sarah put on too many airs, and she resented her ownership of Broadhall. Whereas Herman enjoyed celebrating Sarah as the "Lady of Southmount," Maria scornfully referred to her as "the lady" and "Madame."

What neither George nor Maria would have known was that this dinner ceremony to crown Herman was inspired by a scene in a book. Sarah had made a point of privately sharing that book with Melville earlier in the year. It was a romance with an exotic title—*Zanoni*, by Edward Bulwer-Lytton, a friend of Charles Dickens. In an early chapter—only a dozen or so pages into the story—an opera composer of genius sees his neglected masterpiece performed at last, and at a dinner to celebrate the event, his proud wife "suddenly" steps forward and places "on the artist's temples a laurel wreath, which she had woven beforehand in fond anticipation" of his success. The episode stands out in the story as proof of the composer's triumph not only in art, but also in love. Like *Moby-Dick,* the opera deals in part with the mysteries of the sea. The climatic moment features a Siren queen emerging from her ocean cave.

Earlier in the summer of 1851, when Melville was putting the finishing touches on *Moby-Dick,* Sarah read *Zanoni* and was eager to share it with him. It is the story of a profound romantic attraction between an artistic young woman and a mysterious older man in the days of the French Revolution. Instead of sharing *Zanoni* with Melville in a stolen moment together, Sarah did something more daring. She packed it in a box containing a second book (Harriet Martineau's *The Hour and the Man*), then added "two flasks of Cologne" and sent the whole thing off to Melville as a surprise gift. It was an astonishingly provocative gesture. The typical married woman of the time wouldn't even dare to dream of sending such things to another woman's husband, but, of course, Sarah wasn't typical, and Melville didn't discourage her. He welcomed the gifts as "nourishment for both body & soul," replying with a letter of thanks and calling her "the most considerate of all the delicate roses that diffuse their blessed perfume among men."

This was high praise indeed from a man who considered roses the finest of all flowers, a "voucher of Paradise," as he once called them because they grew so profusely around Eve in the Garden of Eden. His reply contained the promise to read the two books as soon as he could find the time. Meanwhile, he wrote affectionately, "I shall regard them as my Paradise in store, & Mrs. Morewood the goddess from whom it comes."[5]

FOUR DAYS AFTER the Christmas dinner at Broadhall, when Mr. and Mrs. Morewood came to Arrowhead for a visit, Herman wasn't there, and the simmering tensions in the households bubbled over. In Sarah's view, his mother did the unforgivable—she began judging her. Sarah recounted a mild version of their clash to George. "Mrs [Maria] Melville tells me that I have some good in me—she also says that my real character is not yet formed—that I have in fact no fixed purpose in Life—So she judges me—little knowing my real feelings—So you must not judge me—for I mean to acquire *a real decided character*." Such was Maria's imposing demeanor—her "gorgon" look, as D. H. Lawrence would have said—that few in her circle would have dared to stand up to her. But Sarah did.[6]

Afterward, a ruffled Maria gave Augusta the details, warning her not to let Melville's brother Allan see the account.

> Her mind was in an excited state & from all accounts must have been so all day, she express'd ungrateful feelings towards Mrs Brittan [Sarah's older sister], who has done more to make Broadhall comfortable in the past two weeks than the lady has done in the past year. She said many things to her "liege lord" [Rowland] which even his long patient forbearance could not let pass. Altogether so much more was said that I requested her to take some thing to quiet her nerves, as She did not seem to like this, I at last told her that her conversation affected her Husband very painfully & I

wished her to change the subject. She said she felt rebuked, but she must speak out when she felt so full [of emotion] & she could not help it. She also said the Bible was not written by inspiration etc.[7]

The mere words on the pages of Maria's letter can't convey the explosive force of this encounter. Young Dutch girls from New Jersey didn't talk back this way to dignified Maria Melville of the grand old Gansevoort family of Albany, New York. And how dare she have the nerve to say to Maria—good Christian woman as she was—that the Bible wasn't divinely inspired? Or to show disrespect to the man joined to her in holy matrimony as her lord and master?

There was no reason for the Lady of Broadhall to cross swords with Maria unless she felt that more was at stake than simply a neighbor woman's opinion of her. This wasn't about holiday jitters or housekeeping or the Bible. It was Sarah finally lashing out at a narrow-minded, self-centered woman who couldn't recognize what a burden she and her daughters had become for Herman. As far as Sarah could see, they didn't understand or appreciate his genius. They had no idea of his secret life with a woman Maria judged as impious and morally weak. It is a wonder that Sarah didn't blurt out the truth of their affair while she was at it.

Even this brief glimpse into the real Mrs. Morewood—edgy, provocative, passionate—wasn't enough to awaken Maria to the possibility that her son might find such a woman irresistible, and far preferable to Judge Shaw's daughter. For Sarah, however, this turbulent holiday week did seem to have a sobering effect. The tone of her letters to George became less playful, and she made a point of explaining how much she valued Herman's friendship. "He is a pleasant companion at all times," she wrote of Melville, "and I like him very much." This sounds harmless to modern ears, but it was exactly the kind of thing that unnerved George. He knew how dangerous it was for a married woman to speak of any man other than her husband as a "pleasant companion at all times." What he couldn't have known was how much she was hiding even when she appeared to be writing in earnest.

To win George's sympathy—and by extension the goodwill of the *Literary World* journal—she often tried to seem like a conventionally religious woman who was searching for answers to needs and desires that godly men like the Duyckincks might help her to reconcile with her faith. Melville was anything but godly, so Sarah acted as if that aspect of his character concerned her. Of course, as she had revealed to Maria, she was herself breathtakingly irreverent for a woman of her station. "It is a pity," she solemnly informed George, "that Mr Melville so often in conversation uses irreverent language—he will not be popular in society here on that very account—but this will not trouble him—I think he cares very little as to what others may think of him or his books so long as they sell well." If George had been able to see Sarah doing battle with Maria, he would have known that the author of *Moby-Dick* was not the only fearlessly irreverent voice in Pittsfield.[8]

IN FEBRUARY 1852 Melville decided that he couldn't accept what the Duyckincks had done to him in the pages of the *Literary World*. The journal's criticism of *Moby-Dick* was more than he could take. He stopped his subscription with a cold and blunt demand to "discontinue" sending the *Literary World* to "H Melville at Pittsfield." For the next few years he would have nothing to do with the brothers. Accordingly, Sarah's own efforts to win George's favor dwindled to sporadic letters, and they had relatively little contact. Sarah didn't fail Melville, but she couldn't rescue him from the deep hole that he had created for himself. He wouldn't have wanted her to rescue him—even if she could. He was too proud for that, and he was so talented that he continued hoping his art could somehow save him. With *Pierre*, he had one more chance, though it was risky on so many levels, especially given its revealing look at his secret life.[9]

THINGS CONTINUED TO SPIN DOWNWARD. By Christmas he must have known that there was one good reason for Sarah to be a little edgier than usual. Four years after giving birth to her first child she was pregnant again. She had conceived at some point in September, a few weeks after the Greylock excursion. That was the period when her husband was rarely in Pittsfield, and she was spending a lot of time with Melville wandering the beautiful countryside with his spyglass, and giving him romantic books and cologne as presents. On some of those occasions they were joined by others, but the trail connecting their homes made it easy for them to be together whenever they chose during those mild late summer days. Sarah was alone with Rowland for six months in England but didn't become pregnant again until she spent her first full summer at Broadhall with Melville as her neighbor and lover.

It may not be possible to prove anything one way or the other at this late date, but the emotional fireworks over the holidays would make a great deal more sense if Sarah had known then that she was going to have Melville's child. It would also help to explain why she was even willing for a time to take a more charitable view of Herman's family, saying that she was "strangely and strongly attracted" to them. It would explain why, on Christmas Day, Melville made a point of leading Sarah into dinner as if they were the master and mistress of Broadhall. Most of all, it would help to explain why the often charming love story in the first half of *Pierre* turns so dark and desperate and guilt-ridden in the second half of the book. In fact, the novel became what the Harvard critic F. O. Matthiessen called "about the most desperate in our literature." (And Francis Otto Matthiessen knew something about desperation. One morning in 1950 he ended his life with a twelfth-floor leap from the window of a Boston hotel.)[10]

Sarah would seem to have been prepared for what used to be called a love child. When the writer Margaret Fuller (whose full name was Sarah Margaret Fuller) returned to America in 1850 from a long stay in Italy, carrying an infant that many thought was born out of wedlock, her ship ran aground off Long Island, and she and her child were drowned. The whole episode made a lasting impression on Sarah. At the end of the month in which she became pregnant, she wrote, "I remember reading with pain an account of the shipwreck—and my thoughts at the time that it was one of life's deep romances." It was odd to think of romance in response to such a tragedy, but perhaps not for Sarah as the crucial month of September 1851 was coming to an end.[11]

Proper American ladies took a different, and much harsher, view of that famous shipwreck. Hearing the news of Fuller's drowning, the otherwise kindly Sophia Hawthorne thought that, given the questionable circumstances of the baby's birth, and the character of the young Italian man Margaret had brought home as her new husband, the lady was better off dead. (The husband also shared that sad fate.) "I am really glad she died," wrote the wife of the author of *The Scarlet Letter.* "[T]here was no other peace or rest to be found for her—especially if her husband was a person so wanting in force & availability."[12]

Yes, Professor Matthiessen grasped exactly how desperate Melville's mood turned in the closing chapters of his most autobiographical work. Even if Sarah's child wasn't his, the suspicion was enough to make Maria Melville's son dwell on some very dark thoughts indeed in the coming winter months. The worst thought wasn't any worry about how Pittsfield might react to his secret life. It was what Beacon Hill might say that was most troubling. Judge Shaw's grim, jowly visage must have floated many times over Melville's view of Greylock as he sat at his desk. After giving the judge two grandsons, how would he explain Mrs. Morewood if the truth came out? It was a good thing he had experience jumping ship in the middle of the Pacific. If he shamed the judge's good name in Boston, there wouldn't be any place too remote for him to hide.

PART III
—

THE VOYAGE OUT

She vanished, leaving fragrant breath
And warmth and chill of wedded life and death.

—HERMAN MELVILLE

19

"STEEL, FLINT & ASBESTOS"

Under the headline A FATAL OCCURRENCE, a New York satirical paper—the *Lantern*—published a bogus news item in 1852 about the dangers of reading Herman Melville. *Pierre* had just been published in the summer by Harper & Brothers and reviewers agreed that Mr. Melville had lost his mind. The joke behind the little news story in the *Lantern* was that the author's supposed madness was so consuming it might be infectious: "About ten o'clock yesterday, an intelligent young man was observed to enter the store of Stringer and Townsend, the well-known publishers, and deliberately purchase a copy of Herman Melville's last work. He has, of course, not since been heard of."[1]

That's it, that's all there was to the story. And that's all there needed to be. The reception of the book was so bad that the *New York Evening Mirror* said Melville should feel ashamed for having written it. The stylized, archaic dialogue, the "unwholesome" theme of incest, and the vague complications of the love story left one reviewer feeling that Melville was suffering from a bad case of food poisoning. The book was "one long brain-muddling, soul-bewildering ambiguity . . . the dream of a distempered stomach, disordered by a hasty supper on half-cooked pork chops." Whatever malady was troubling Melville, said a New York critic, the author had succeeded in accomplishing at least one thing with his latest work: "The highest literary reputation ever achieved would be demolished by the publication of a few volumes of such trash as this *Pierre*—a novel, the plot of which is monstrous, the characters unnatural, and the style a kind of prose run mad."[2]

The idea that Melville must have suffered some sort of mental collapse soon caught on, and in September a New York headline appeared with the words "Herman Melville Crazy." The article made the false charge "that Melville was really supposed to be deranged, and that his friends were taking measures to place him under treatment. We hope one of the earliest precautions will be to keep him stringently secluded from pen and ink." It didn't help that Melville decided to dedicate the book to "Greylock's Most Excellent Majesty." Some reviewers assumed that anyone dedicating a novel to a mountain must be crazy. It would have been more appropriate, said a southern newspaper, "if he had dedicated it to the Lunatic Asylum."[3]

The reviews were especially harsh in Boston, which must have been profoundly humiliating for Judge Shaw and his family. "Utter trash," said the *Boston Post* in August, calling the novel "the craziest fiction extant." The *Boston Daily Times* proclaimed that Melville's worst passages were some "of the absurdest and most ridiculous things that ever ink and paper were wasted on." And the city's *Evening Traveller* said the novel was not only "extremely disagreeable," but also "unnatural and improbable." The criticism that Melville must be insane started, in fact, in Boston with the *Post*'s review. The last part of that article was a devastatingly personal attack on the man of Pittsfield: "What the book means, we know not. . . . It might be supposed to emanate from a lunatic hospital rather than from the quiet retreats of Berkshire. We say it with grief—it is too bad for Mr. Melville to abuse his really fine talents as he does. A hundred times better if he kept them in a napkin all his natural life. A thousand times better, had he dropped authorship with *Typee*. . . . As it is, he has produced more and sadder trash than any other man of undoubted ability among us."[4]

The character Pierre's publisher—a firm like Harper & Brothers that Melville's caustic wit turns into "Steel, Flint & Asbestos"—accused the young writer of wasting the company's time with a book that is "a blasphemous rhapsody." Isabel and Lucy pull at his heart until he feels that he is about to disappear into "a black, bottomless gulf of guilt." And, like Ahab chasing the whale, Pierre sees himself standing on the quarterdeck of a grand vessel on the verge of cracking up: "His soul's ship foresaw the inevitable rocks, but resolved to sail on, and make a courageous wreck."[5]

All the usual qualifications can be endlessly rehearsed for fiction as something that can't be reduced to mere autobiography. Melville may have wanted to use fiction to transcend the painful reality of his life, but the problem that sank *Pierre* was that the novelist raised the white flag and surrendered any pretense that he was writing fiction. Though his contemporaries still couldn't understand the exact nature of his problem because he couldn't give specifics, they quickly jumped to the conclusion that it was deeply personal, throwing out charges of insanity almost from the start. Modern critics have been able to see that Melville was laboring to reveal something about him-self and his recent struggles, and have tried to guess the problem, but have never sought to link it with Sarah. About two-thirds of the way into his strange love story, all the romance of a young man falling for a mysterious beauty crumbles under the weight of an author trying to explain what is really happening to his life. In essence, he admits that no novel can do justice to his story. The incest plot wears thin, the Gothic trappings fall away, and what is left at the end is a young au-thor with a story he longs to tell, but can't. The truth would hurt too many people, the passion he feels for his forbidden lover would cause a scandal, and his remaining friends would turn on him. So Melville's hero folds personal failures into the larger context of a corrupt civili-zation and an indifferent cosmos, and in time young Pierre begins to sound a lot like old Ahab, cursing creation and lamenting "the ever-lasting elusiveness of Truth."

IN *PIERRE*, Melville comes face-to-face with the darkest questions of the last two years. What was the good of finding love when you can't have it? What was the good of writing a brilliant book that no one will read? As Pierre sits in his "shivering room," he thinks that his efforts at greatness have only been self-defeating. "At last the idea obtruded, that the wiser and the profounder he should grow, the more and the more he lessened the chances for bread, that could he now hurl his deep book out of the window, and fall to on some shallow nothing of a novel . . . then could he reasonably hope for both appre-ciation and cash."[6]

Like Melville, Pierre is ready to endure a grueling routine in soli-tude to produce a masterpiece. In the isolation of his room the only sound is "the long lonely scratch of his pen." *Pierre* raises a troubling thought taken straight from Melville's own bitter response to the failure of *Moby-Dick*. "In the heart of such silence, surely something is at work. Is it creation, or destruction? Builds Pierre the noble world of a new book? Or does the Pale Haggardness unbuild the lungs and the life in him?—Unutterable, that a man should be thus!" Profoundly discouraged, Pierre confronts the hard fact that he can't create the book he wants because the world prefers pretty lies to harsh truths. Perhaps the most moving lines in all of Melville's work are those in which Pierre admits that what he is most eager to tell will never be written in ink. "Two books are being writ," he confesses, "of which the world shall only see one, and that the bungled one. The larger book, and the infinitely better, is for Pierre's own private shelf. That it is, whose unfathomable cravings drink his blood; the other only demands his ink."[7]

In the process of writing his novel, which stretched from the late summer of 1851 to the spring of 1852, Melville—like his protagonist—watched his world in the Berkshires go from an idyllic dream of love and nature to a nightmare of domestic turmoil, professional failure, crippling debt, possible scandal, and emotional despair. When he finished the book in April, Sarah was seven months pregnant, and he must have assumed the world would soon collapse around him. The truth of his affair would come out, his family would disown him, and Lizzie's family would hound him into the grave.

That is the kind of scenario he creates for Pierre, whose mother disinherits him after he runs off with Isabel, and whose life is threatened by Lucy's brother. So strong was Melville's sense of impending doom that he paused near the end of his novel to write a searing epitaph for anyone anxiously awaiting a moment of public disgrace.

"Not the gibbering of ghosts in any old haunted house; no sulphurous and portentous sign at night beheld in heaven, will so make the hair to stand, as when a proud and honorable man is revolving in his soul the possibilities of some gross public and corporeal disgrace. It is not fear; it is a pride-horror, which is more terrible than any fear."[8]

Readers of *Pierre* are always surprised when suddenly Melville invests his hero with a recently acquired literary celebrity on the basis of some highly acclaimed verses. When he is falling in love, Pierre is just a carefree youth, but when romance becomes complicated and fraught with danger, he acquires as if from the blue a full-fledged literary career. He must have one—however implausible it may seem—because Melville can't get close enough to describing his own predicament without it. So the bedazzled lover in the Berkshires, the hero of the early chapters of the book, becomes in the later chapters an increasingly embattled author like Melville, whose best days are behind him, and who yearns to explain himself to a world preparing to bury him with recriminations and ridicule. But the question of a "public disgrace" awaiting Pierre never seems a real threat for his literary sins. The only thing that would cause such a debacle would be the revelation of his love for a woman claiming to be his half sister. "Unnatural" is the condemnation that would destroy him. His public profile isn't so great, and his vice is too unmentionable, for the press to trumpet his fall.

It was different in Melville's own case. More than any other passage in his book the one on disgrace sounds like the novelist speaking for himself. The degree of public humiliation he was facing couldn't be exaggerated. It would have to be the stuff of large and scathing headlines about Mr. Typee, the judge's daughter, and the rich man's wife. As it was, in the dark isolation of his fears and guilt, living with the household of women he was betraying, and seeing the woman he loved pregnant, he decided to turn his book into a note declaring his professional suicide. If he had to fail, the failure would be monumental and complete.

HE WENT AFTER THE HARPERS, mocking their literary pretensions and their tight-fisted greed in the unforgettable "Steel, Flint & Asbestos," and he took aim at his critics in general and the Duyckincks in particular. As revenge against Evert calling the views in *Moby-Dick* "piratical," he satirized the *Literary World* as the "Captain Kidd Monthly." As for the plot itself, he sent it into a maelstrom of destruction, creating a final death scene like that in an Elizabethan tragedy, with the bodies of the dead lovers crumpled in the agony of their heartbreak and suicide. The scene fades out with Isabel's body slumped over Pierre's, and the final image is a bower of death. "Her long hair ran over him, and arbored him in ebon vines."[9]

When the novel was published in midsummer 1852, the Duyckincks were so appalled by *Pierre* that they questioned whether Herman had been abducted by some devil using "necromancy" and replaced by a pale ghost of the novelist who wrote *Typee*. Now the "piratical" Ahab and crew seemed almost tame against the "leering demoniacal" face of Pierre. Though the brothers knew that Melville's private life was questionable, they weren't prepared for the ferocity of the new novel, and "the stagnant pool" of its moral view, as it seemed to them.

In the *Literary World*'s review they treated the man who had been their host in the Berkshires as a kind of literary outlaw writing books that attacked the foundations of everything sacred. Either he had gone insane or he was now one of the most dangerous writers in America. "The most immoral *moral* of the story," the review charged, "if it has any moral at all, seems to be the impracticability of virtue. . . . Ordinary novel readers will never unkennel this loathsome suggestion." Though they couldn't understand the reason why, they were right about this "suggestion." If society tells Pierre that loving Isabel is wrong, yet it seems the best thing that ever happened to him, how is he to love her and be virtuous? The novel doesn't really treat the "impracticability of virtue" as a suggestion. It treats it as a fact, and the Duyckincks couldn't fathom such a thing, no matter how sweetly Sarah wrote to George or how generous Herman was to Evert.[10]

Pierre's world is one that Melville had hoped Hawthorne would help him explore—a puritanical sphere aflame with sin and punishment, where good women like Hester or more daring women like Sarah can't be anything but bad. What Melville had discovered was that although Hawthorne knew some of the mysteries of that sphere, he didn't know what it was like to be immersed in it. Herman was immersed in it, right up to his neck, and *Pierre* is his agonizing view of the guilt and fear infecting what started out as simply a pleasure.

While Melville was hard at work on the book, Sarah had gently teased him about the strain on his mind, and he had lightly dismissed the subject: "I laughed at him somewhat and told him that the recluse life he was leading made his city friends think that he was slightly insane—he replied that long ago he came to the same conclusion himself." But the disaster of *Pierre* was no joke. Under a barrage of negative publicity, the book simply stopped selling. Over a period of six months after publication Harper & Brothers would manage to sell fewer than three hundred copies.[11]

This was the catastrophe that Melville dreaded, the "courageous wreck" of a career and a life that he had come to view with such bitter disillusionment. And, of course, he had brought much of it on himself, out of anger and frustration. To outsiders, the collapse was inexplicable. The only answer was that the author must have lost his mind. But he didn't lose his mind—he lost his nerve. He was young, impulsive, and still something of the "proud and savage" sailor who had taken his chances on the run as a castaway in the Pacific. This time, when he went overboard, he had good reason to expect that he wouldn't resurface.

HE JUMPED TOO SOON. Sarah had her baby in June—a boy she named Alfred. Her husband didn't make a fuss, and must have assumed the child was his. The women at Arrowhead didn't seem any the wiser, and there was no general outbreak of gossip beyond the knowing glances of astute observers like Dr. Holmes. Alfred grew up at Broadhall, and though he bore a little resemblance to Melville's sons by Lizzie, it wasn't close enough to stir suspicion.

Instead of a shattering explosion that summer there was only a long sigh of regret from friends and admirers who couldn't understand what had driven their young novelist over the edge. Readers have been wondering the same thing ever since. "How is one to account for the transformation of this apparently normal young man into the savage pessimist who wrote *Pierre?*" asked the novelist Somerset Maugham in the twentieth century. "What turned the commonplace undistinguished writer of *Typee* into the darkly imaginative, powerful, inspired and eloquent author of *Moby-Dick?*"

The short answer is falling in love. Maugham guessed as much, though he didn't have Sarah's name, and didn't know how or why the author strayed from his wife. "I think it is probable," he conjectured with old-fashioned politeness, "that Melville was impatient with the marriage tie; it may be that his wife gave him less than he had hoped." Well, yes, a lot "less." Mountains' worth of "less." And that's a clue that there was some method to what critics thought was madness in *Pierre*.[12]

Hovering over the book is the ghostly image of Mount Greylock, which is known in Pierre's world as the Mount of Titans. It is the same distance from his ancestral home as Greylock is from Broadhall and Arrowhead—fifteen miles—and there is a dark, mysterious place on its slopes that Melville describes in words almost identical to Sarah's recollection of her favorite place on Greylock. "The wild wood circle I shall never forget," she says in her *Taghconic* memoir of the August excursion with Melville and their friends. It is "formed by fir trees so densely shaded with thick foliage as to exclude a single peep from the bright face of Sol; while the grass growing was of a light, moss color, of that peculiar green seldom to be found, except in small tufts by a shady brook side. And then the silence and repose of the place had the effect of awing one, as it were, and making one superstitious, in spite of oneself."[13]

Near the end of *Pierre*, Melville writes of a similar bower on the mountain slopes as a mossy spot with "a hidden life" so perfectly protected from view that it is cloaked in darkness even at midday in August. "Now you stood and shivered in that twilight, though it were high noon and burning August down the meads." But, as Melville's use of "shivered" suggests, its darkness has acquired a chill. In *Pierre* he looks back at the spot as a place whose charm has turned gloomy. Its beauty now torments him because it is so empty and desolate. The "happy hours" have fled and all that remains is "ruin, merciless and ceaseless; chills and gloom."[14]

Once readers understand the influence of Sarah and Greylock in the story, it is easier to understand Melville's seemingly strange choice for the dedication page of *Pierre*: "To Greylock's Most Excellent Majesty." It should be obvious now that Melville saw Greylock as a silent witness to the twin dramas of his recent life—writing *Moby-Dick* and falling in love with Sarah—and that both these dramas helped to inspire *Pierre*. It is also possible that he wasn't recognizing the mountain so much as the woman. The dedication has always been taken as a reference to the grandeur of Greylock, but it can also be read as a tribute to the reigning spirit of the Berkshires in Melville's eyes—the regal Lady of his letters, the goddess in his Berkshire paradise.

The book should have been dedicated to Sarah by name, but to have done so would have caused a firestorm in Pittsfield. Instead, cleverly, his phrase can be read two ways, and only Melville and Sarah would have understood how he was honoring her—his "ever-excellent & beautiful Lady of Paradise"—as the mountain's queen, "Greylock's Most Excellent Majesty." In *Pierre; or, The Ambiguities,* an ambiguously worded dedication makes the perfect start. But, of course, the literal-minded critics of Melville's day didn't want ambiguity. That is a modern taste. They wanted an adventure from him that needed no interpretation. When they dipped into *Pierre* and found ambiguous lovers saying ambiguous things, they threw their hands in the air and gave up, calling the book unreadable and worse.

SARAH WOULD NEVER let Melville forget the trip to Greylock. A few years after *Pierre* came out, when things weren't going well in life for either of them, she would send him a beautiful copy of Edward Bulwer-Lytton's *The Pilgrims of the Rhine,* bound in expensive leather with gilt trim and many elaborate engravings. It was a sumptuous present to give to any friend, but it had a special significance related to Greylock, a reminder of a shared delight in a particular spot on the mountain. As Sarah pointed out in her *Taghconic* essay, her enchanted place on the slopes was so green, peaceful, and secluded that it reminded her of scenes in one of her favorite books, *The Pilgrims of the Rhine*. She would write in Melville's copy the simple inscription, "Herman Melville From his Friend S.A. Morewood Jan. 1st 1854." The title page features a short quotation from Percy Bysshe Shelley that might have conveyed what she really meant to say. It begins, "Wilt thou forget the happy hours / Which we buried in love's sweet bowers?"[15]

In every practical sense, the author of *Moby-Dick* saw his literary career come to an end around August 1852, when he was only thirty-three. If he had taken his life that summer, as he may well have been tempted to do, the full story of his tragic rise and fall might have been revealed long ago by his saddened friends and family. It is easy to imagine how the world would romanticize a legendary seafarer who survived the perils of the deep only to find disenchantment on land and die young. In such a scenario *Moby-Dick* would surely have won renown much earlier, lauded as the young man's greatest achievement, the book he died for. The laurel wreath might have adorned his tomb long before the end of his century.

After the failure of *Pierre*, the love story of Herman and Sarah continued, though at a more subdued pace, and with a great struggle—as we shall see—to cope with continuing money troubles, and—especially for Sarah—illness.

20

THE COUNTESS

Sarah was right to think that she needed to seize her pleasures from life while she could. She turned thirty in 1853, and for much of the decade ahead she would struggle to stay healthy. In August of that year, at the height of the Berkshire season, Lizzie Melville noted in a detached tone that something wasn't right up at the mansion on the hill. "Our neighbors at BroadHall," she wrote to Judge Shaw, "are neighborly, as usual, though Mrs. M. not being very well this summer, and having a child sick also, is not in her accustomed picnicking mood—so altogether we have had a very quiet summer." The main trouble was in Sarah's lungs. Summers in the Berkshires were usually good for her, but in winters she would have bad coughing spells. What she was fighting used to be called consumption, the old term for the slow, consuming illness of tuberculosis.[1]

In its early stages the disease was thought to have the magical effect of enhancing a woman's appearance, heightening the glow of her face and eyes and giving her a sad radiance as if she were doomed to be consumed by her own beauty. Even when energetic and full of passion, the great consumptive heroines in the literature of the time would show a touch of melancholy in their eyes that men of the period found captivating. Edgar Allan Poe, a devoted admirer of such women, was capable of writing rapturously about "the gentle disease," celebrating what he thought was the slow ecstasy of its progress: "How glorious! To depart in the heyday of the young blood—the heart all passion—the imagination all fire—amid the remembrances of happier days—in the fall of the year—and so be buried up forever in the gorgeous autumnal leaves."[2]

The reality was not so romantic, but in her late thirties Sarah often had the classic haunted look of a languishing beauty. In *Pierre*, Melville plays up the appeal of Isabel as a beauty with the hint of some consuming illness waiting to overtake her, a woman with an "immortal sadness" in her face, and then toward the end of the novel—as mentioned earlier—she makes that poignant lament for her fragile health, "Was ever hearse so plumed?"

On the other hand, there were also times when Sarah seemed as full of life as ever, renewing her social activities with even greater vigor or galloping away on her latest steed, Kossuth, named after the Hungarian revolutionary whose speeches she admired. When Rowland took her to South Carolina for a rest cure in December 1854, she was feeling terrible, yet she still managed on most days to go for rides of ten miles on horseback. Regardless of her health problems, she was strong enough at thirty to give birth to another child, a daughter. Anne Rachel Morewood was born in November 1853, about a year and a half after Sarah's second son, Alfred.

Again, it is impossible to say for certain whether the father was

Rowland or Herman. After the publication of *Pierre*, Melville was so guarded about his private life, and so cut off from the literary world, that it's harder to track his movements. What we do know is that he struggled for years to cling to the unprofitable farm whose only real value for him was its location next to Broadhall. He never stopped loving Sarah, as we will see.

There are two clues about his possible relationship to Anne Rachel. One is that thirty years later he did something highly unusual for him in his often difficult and reclusive old age: he composed an affectionate toast for a wedding celebration. "The Fair Bride" was the opening salute he sent as his greeting to Anne on her wedding day at Broadhall. By that time, he had not lived in the Berkshires for almost twenty years, and wasn't in close touch with Rowland, then a widower. There was no apparent reason for him to acknowledge Anne's marriage, but he did so with these wistful words: "Wherever fortune carries her may she remember Berkshire, beautiful and happy, as Berkshire will always remember her."[3]

The second clue is in the immediate aftermath of Anne's birth, in December 1853, when Melville gave a curious twist to his usual habit of bestowing fanciful names on Sarah. He sent her a letter marked "Particularly Private and Exclusively Confidential" and addressed it, "For, The Honorable & Beautiful Lady, The Countess of Hahn-Hahn—Now at her Castle of Southmount." As improbable as it may seem, there really was a Countess Ida von Hahn-Hahn in Germany, and no romantic young man would use that as a nickname for his lover unless they both shared the same racy sense of humor.[4]

In 1852 the *Christian Times* had singled out the countess as an especially undesirable figure whose fervid romances were some of the "most licentious and scandalous" in Europe. Only one of her novels—*The Countess Faustina*—had been translated into English, but that was one too many for the guardians of public morality. The heroine ("one of those souls of fire, who desire continually to be draining the cup of happiness") has been described as "a female Don Juan" in a novel where "adultery is glorified." Worried about censorship, the English publisher suggested to readers that they take Faustina "not as an example but as a warning." When it was reported in 1851 that Countess Hahn-Hahn had suddenly reformed and become a Catholic, the *New-York Tribune* scoffed at the news and said that no one would believe it until she chose to "burn some of her books rather than reprint them."

Given their usual banter, Sarah must have enjoyed Melville's mischievous suggestion that she and the scandalous countess were twin spirits. But if the newborn Anne wasn't his daughter, he was far overstepping his bounds to close with his "compliments" to "that sweet heiress of your noble name, the infant Countess Hahn-Hahn." It was one thing to joke about adultery with Sarah, but why bring her husband's newborn into the joke unless, of course, Anne wasn't Rowland's, but his?[5]

AS THE YEARS WENT BY, Herman and Sarah seem to have accepted that they would never live together, and would always have to keep their relationship a secret. They promoted the notion that they were just close neighbors, bookish friends, and often acted as if the Morewoods and the Melvilles were one big family. Rowland and Herman became friendly, as did Lizzie and Sarah.

Unlike her neighbor, Lizzie didn't dwell too much on romantic subjects, and didn't take literature too seriously. If Sarah's head was often in the clouds, Lizzie's was preoccupied with practical matters involving her family. By the mid-1850s, she was the mother of four children, two boys and two girls. (Elizabeth was born in May 1853, and Frances in March 1855.) Unlike Sarah's businessman husband who was so busy in New York, Lizzie's husband was constantly struggling to provide for his family. If her father had not supported them so generously over the years, they would have surely lost everything.

Gradually, Melville's sisters and mother moved away, except for Augusta, who always stayed close to her brother. Of all his relatives, Melville's mother never warmed to Sarah. In the next generation, relations became so close that William Morewood—Sarah's oldest child born before she met Herman—married Melville's niece Milie. She was his brother Allan's oldest daughter, and her full name was the same as her grandmother's. When she married in 1874, Herman must have savored the irony that there was now a Maria Gansevoort Morewood in the family.

Melville's mother had so little understanding of his talent that she was ready to see him abandon a career that seemed nothing but trouble to her. In 1853 she wrote to one of her relatives that Herman needed a "change of occupation." The profession of writing was draining her son's strength and doing nothing to help the family. Whether Herman knew it or not, the enforced isolation of the writer's life "does not agree with him," his mother concluded. "The constant working of the brain, & excitement of the imagination is wearing Herman out," she complained. No doubt Melville often heard such remarks from her, and with each reminder of his failures he must have found his struggles all the more difficult to bear.[6]

Financial pressures over his debts and his lack of income shadowed every year that Melville remained in Pittsfield. He was often elsewhere, and he found it harder to keep in touch with Sarah. When he was visiting the Shaws in Boston with Lizzie one year, he wanted to alert Sarah that he would be coming home to Arrowhead soon. Yet it was awkward to get a letter to her without the rest of the Shaw household noticing. Arranging to travel ahead by himself to prepare Arrowhead for Lizzie's return, he sent a friendly notice to Mrs. More-wood that he was on his way.

Written with the knowledge that Lizzie would probably see it, this note to Sarah is completely different from his other surviving letters to her, beginning with his uncharacteristically formal salutation, "My Dear Mrs. Morewood." There are no references to goddesses or paradise or knights or the Countess Hahn-Hahn in this letter. It sounds more like a business message to an elderly aunt than one to a woman with whom he had shared so much of importance in his life. With stiff decorum, he signed it, "Very Truly & Sincerely Your Friend & Neighbor, H Melville." He couldn't have done otherwise, for Lizzie did indeed inspect it and added a message of her own to it, supplying some church gossip and a playful but condescending remark about Herman doing as he had been told: "Wives propose—husbands dispose—don't you think so?" she asked Sarah.

From Melville's point of view the only real purpose of this letter was to let Sarah know that he was coming home and would be alone. She had written earlier to Lizzie, suggesting that the family could stay at Broadhall a few days when they returned if they wanted. But Melville was obviously looking forward to enjoying his own time with her, and had led Lizzie to think that his early return was all her idea. No doubt with a secret sense of excitement he was able to announce to Sarah in an open and seemingly innocent way

that he would "be happy to accept, for myself, your kind & neighborly invitation for a day or two."[7]

In this way, Sarah and Herman turned an invitation for all into
an invitation for one, giving themselves a precious couple of days on
their own at Broadhall, with servants silently looking the other way,
and neighbors gossiping as usual about Mrs. Morewood's strange
manner of hospitality. The formal letter made it all seem so proper.
In such covert ways, the affair was able to continue long after the
initial romantic upheaval had settled into something less dramatic. It
was the best they could manage under the circumstances. Melville had
realized early on that such a path might stretch out ahead of them,
writing of Pierre's ongoing relationship with Isabel that it was full
of "the secretness, yet the always present domesticness of our love."[8]

Yet Melville had wanted so much more from this affair, and he
bitterly regretted that his setbacks had taken from him the chance to
turn the Berkshires into the perfect retreat for a successful novelist
and lover. When at some point in the late 1850s he gave Sarah his
copy of Dryden's poems, he marked not only the line about the first
night of love for Sigismonda and Guiscardo, but also a passage about
the tragic end of their hidden love.

Discovering that the princess has secretly wed Guiscardo, her
jealous father has the man killed and sends the heart in a golden goblet to Sigismonda. So devoted is the princess to her dead lover that she
immediately resolves to poison herself rather than live without him,
and presses her lips to his heart in a last kiss.

It is an extraordinary moment of distraught passion that Melville
may have first encountered in an engraving of William Hogarth's
painting of an overwhelmed Sigismonda clasping the goblet to her
breast. In his gift of the Dryden volume to Sarah, Herman made a

long, dark penciled line down the margin beside the description of the princess kissing her lover's heart. Because he presented the poetry volume to her in the later years of their relationship, it is easy to see that it gave Sarah two passages that are like bookends to a romance, one signifying a bright beginning, the other a sad fate imposed by outside forces. (Like Sigismonda, Isabel and Pierre die by poison in Melville's novel.) These high and low points of love give an overview of the personal journey Melville undertook, largely in secret, in his thirties and early forties. It would take the rest of his life for him to come to terms with it.

21

ROUGH PASSAGE

On a cold, rainy Saturday in November—one of those bleak occasions that Ishmael calls "a damp, drizzly November in my soul"—Herman Melville was warming himself in the snug comfort of the White Bear Hotel in Liverpool, England. It was 1856—five years after *Moby-Dick*'s publication—and the author was at the start of a long trip, largely financed by his indulgent father-in-law the chief justice, who had reacted more in sorrow than anger to Herman's failures.

Earlier in the year Melville had narrowly avoided financial ruin when he paid off some of his debts by selling half of his farm acreage, and by securing another large loan—five thousand dollars—from Judge Shaw. He had little money, few prospects, and in recent months his health had been bad. He had overworked himself trying to salvage something from the wreck of *Pierre*. Book publishers were now reluctant to deal generously with him, so he had been concentrating lately on shorter works of fiction for periodicals, producing a handful of brilliant works, including "Bartleby, the Scrivener" and "Benito Cereno." He also wrote a novel that no one published—"The Isle of the Cross"—and two novels that their publishers probably wished they hadn't published—the historical fiction *Israel Potter,* and the dark, almost impenetrable satire *The Confidence-Man.*

Nothing he did made any difference to his fortunes. His reputation as a novelist was well and truly ruined. Poetry became an option because he didn't have any reputation as a poet, and he could write for himself if he chose and not bother with publishers. As far as full-scale novels were concerned, he was done with them for life. It is tempting to imagine him as another veteran of the Dead Letter Office, like his Bartleby, who will resolutely reply, "I prefer not to," if ever asked to write another book as large and majestic as *Moby-Dick.*

In her letters to her family on Beacon Hill, Lizzie had told her father of her worries for Herman's health, especially his mental state, and it was at her urging that the judge had agreed to help him go abroad for at least several months. "I think he needs such a change," her father said, "& that it would be highly beneficial to him & probably restore him." As he explained to one of his sons, "I suppose you have been informed by some of the family how very ill Herman has been. . . . When he is deeply engaged in one of his literary works, he confines him[self] to hard study many hours in the day, with little or no exercise, & this specially in winter for a great many days together. He probably thus overworks himself & brings on severe nervous affections." It was clear that both his health and his career were in decline, and it was only natural that his family thought a long voyage would help. Even the local newspaper gave its approval to the idea. "Mr. Melville much needs this relaxation from his severe literary labors of several years past," said the *Berkshire County Eagle,* "and we doubt not that he will return with renovated health and a new store of those observations of travel which he works so charmingly."[1]

Beneath the surface there seems to have been a growing fear that Herman would never recover from whatever it was that had made him so restless and unhappy. In old age Lizzie confided to a niece that this overseas trip was in fact "a tentative separation," and—as early Melville scholars discovered—"some members of the family hoped he would get lost and never return." The only bright spot in his life was Sarah. Whenever he wrote to her in private, his prose once again came to life and showed spirit. One of the last letters he sent before going abroad ends with the same light air of his early letters, but with a poignant "adieu" to mark his departure: "So Adieu to Thee Thou Lady of All Delight; even Thou, The peerless Lady of Broadhall. H. M."[2]

HE LEFT HOME in October 1856 and was expected to be away for six or seven months, traveling through Europe and as far as the eastern Mediterranean, including a visit to Jerusalem. There was no real necessity to stop in Liverpool, except for one sentimental reason— the chance to see Nathaniel Hawthorne, now the American consul to the city. Thanks to an appointment by Hawthorne's college friend, President Franklin Pierce, the author of *The Scarlet Letter* was enjoying a long stay in England as one of his country's minor diplomats.

Melville was then thirty-seven, and was still struggling to understand how his great gamble with *Moby-Dick* had blown up in his face. Ahead of him lay more than thirty years in which he would have to bear the burden of that failure. He was so out of touch with even his recent past that he didn't have a current home address for Hawthorne in Liverpool, and was forced to seek him out during office hours at the consulate. When he arrived there on a Monday in early November, he seemed to Hawthorne "a little paler, and perhaps a little sadder." After spending a couple of days together, Hawthorne concluded that his former neighbor had endured a great many hard blows, and the effects showed. "The spirit of adventure is gone out of him," he remarked. "He certainly is much overshadowed since I saw him last." Even Hawthorne's young son Julian, who had found Melville a playful friend in the Berkshires, was struck by the change in the man, who "seemed depressed and aimless."[3]

Like most of those who knew Melville, Hawthorne thought this long trip was a good idea. "I do not wonder that he found it necessary to take an airing through the world after so many years of toilsome pen-labor and domestic life, following upon so wild and adventurous a youth as his was." Though they had not seen each other for a few years, they soon found themselves back in the old familiar relationship of Melville pouring out his thoughts and Hawthorne doing his best to respond sympathetically. As usual, it was an intense experience, but this time—with the cloud of failure hanging so heavy over Melville—Hawthorne couldn't help admiring his old friend's courage in the face of disaster. After a long walk together by the sea one day, talking and smoking cigars, Hawthorne commented in his journal about his friend's strong character: "He has a very high and noble nature, and better worth immortality than most of us."[4]

After about a week in Liverpool, Melville was ready to move on. He would have liked to have spent more time with Sophia Hawthorne, but she was ill during much of his visit. ("Mrs. Hawthorne not in good health," he noted in his journal.) He always valued her kindness, and had been "amazed" when she wrote to him at the time of *Moby-Dick*'s publication, praising the novel in a "highly flattering" way, as he put it. What he especially valued in her, he had said at the time, was her "spiritualizing nature." It enabled her to "see more things than other people."[5]

When Melville said goodbye to the family, Hawthorne was surprised to see that the only thing Melville was carrying was a simple carpetbag: "This is the next best thing to going naked; and as he wears his beard and moustache, and so needs no dressing-case—nothing but a tooth-brush—I do not know a more independent personage. He learned his travelling habits by drifting about, all over the South Pacific, with no other clothes or equipage than a red flannel shirt and a pair of duck trousers. Yet we seldom see men of less criticizable manners than he." The man whose career had once seemed on the verge of shooting straight to the top of American literature wandered off to board his steamer looking like the most ordinary vagabond. For Hawthorne, now the distinguished diplomat and man of letters, the sight of Melville going off to face the world without friends or family or a cartload of baggage was a revelation.[6]

In that parting image of Melville, there is an insight into the castaway mentality that would help the old sailor survive the next few decades. As the most "independent personage" Hawthorne ever knew, Melville was just the type to stay afloat until the right wave washed him ashore somewhere in the world. Whether he was wandering the Berkshires or Manhattan or the Liverpool docks, there was always something of the mariner in him, the surprisingly brilliant stranger whose heart is always in the foretop, searching real or imagined horizons with a magical spyglass and waiting for the next "wonder-world" to come into view.

Invariably, he was disappointed and would spend most of his

later life as a lost soul on the crowded island of mercantile Manhattan, among its robber barons and Wall Street pirates. He would survive until old age claimed him, an ancient Ishmael with a long gray beard, and he would even leave a message in a bottle, so to speak. For at the very end, there was a miracle, a resurrection of the writer still lingering in the old man's soul. He would write one last great book and leave it behind to bob on the waves until someone retrieved it almost thirty years after the old castaway died. It was the manuscript of *Billy Budd*.

Hawthorne would prove less robust, dying at fifty-nine, only seven and a half years after he watched Melville leave Liverpool in November 1856. They would see each other only once more after that day—in the spring of 1857, when Melville was on his way back home from the Mediterranean. But he was in a hurry then, and the only record of their meeting is the short note in his journal, "Saw Hawthorne." As a friendship, it was never entirely satisfying to either man, but for Melville it was really his only chance to share the life of his mind with a true literary peer.[7]

ON HIS SHORT TRIP through England in 1857, Melville stopped for a few days in London. But, of course, no one made a fuss over his visit. He was no longer the literary darling to be wined and dined on the chance that he might write a bestseller. Nobody on either side of the Atlantic believed he was going to do that.

There wasn't even the opportunity for Melville on his quick swing through London to pay his respects to Samuel Rogers. The poet of St. James's Place—friend of Turner and Melville's guide to the painter's art—had finally surrendered peacefully to death in December 1855.

The next year his house had been emptied of all its fabulous contents, and everything was sold off for the remarkable sum of fifty thousand pounds. If Melville had tried to get a peek into the house, he would have found it a ghostly shell, as if the treasures he had seen there were never real, but only part of a dream.

The long trip did little to change Melville's views of his life, and it made no difference in his marriage. He returned to Arrowhead with his family, and he muddled through the next few years, discontented but persevering. A couple of visiting college students who made a "literary pilgrimage" to meet Melville vividly captured his mood in the last years of the 1850s. The students wanted Melville to talk about the South Seas, but the subject didn't hold much interest for him. He was more interested in the failures of the world around him. "He was evidently a disappointed man," recalled one of the students, "soured by criticism and disgusted with the civilized world." The other young man gathered from the conversation that Pittsfield itself was a thorn in the novelist's side. Melville had long ago worn out his welcome and was barely tolerated in the community. "With his liberal views he is apparently considered by the good people of Pittsfield as little better than a cannibal or a 'beach-comber.' "[8]

Knowing hardly anything of the author's personal history, the students wondered why Melville bothered with Pittsfield at all. How could a man who had once explored the balmy regions of the earth resign himself to a lonely life in a cold climate? They didn't dare ask that question, but they came away mystified by his determination "to shut himself up in this cold North as a cloistered thinker." They had no way to know that his heart was still anchored there, despite his disappointments.

Herman Melville

Sarah Morewood

Broadhall in 1900

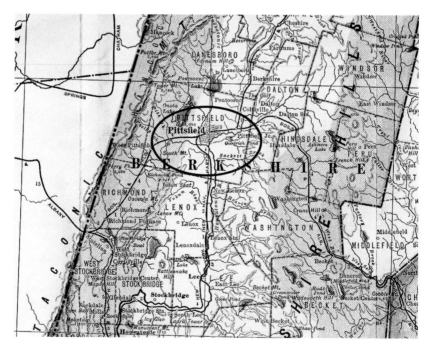

Berkshire County, Massachusetts, with Pittsfield in the center, Lenox
below, and Lake Pontoosuc and Greylock above

Nathaniel Hawthorne

Dr. Oliver Wendell Holmes Sr.

A drawing of Arrowhead by Herman Melville

The view from Melville's room at Arrowhead in 2015

Mary Butler, the bride Melville
"kidnapped" in 1850

The copy of Dryden's poems that
Melville gave to Sarah Morewood

The agreement between Melville and Harper & Brothers for
The Whale, with the last-minute addition of its final title

At the summit of Mount Greylock

Evert Duyckinck

George Duyckinck

Rev. John Todd, Pittsfield's most famous defender of moral values

The children of Herman and Lizzie Melville (*from left to right:* Malcolm, Elizabeth, Frances, and Stanwix)

Sarah Morewood in her mid-thirties with her children (*from left to right:* Alfred, William, and Anne Rachel)

Lizzie and Herman Melville in their later years

22

ASPECTS OF THE WAR

When the Civil War broke out in 1861, Sarah found the great public purpose of her life. Despite her weakening health, she quickly established herself as the leader of the local women helping the troops who streamed in and out of town on training missions and recruitment drives. She organized efforts to give medical supplies and food to the men, held dances and parties at Broadhall for officers, served refreshments to regiments on the march, made flags and various keepsakes for them, and welcomed them home when they returned from the battlefront. As one of her friends would recall, "How her library table was strewn with the photographs and grateful letters of soldiers who had been strangers to her till the war began."[1]

Army camps were named after her in and around two major cities—New Orleans and Washington, D.C. One Massachusetts regiment reported from their camp that the assembled troops raised the silk flags she had presented to them and cheered Mrs. Morewood's name just before they sang the national anthem. One American flag she gave to a regiment came with a staff cut from wood on Mount Greylock. After the war a Massachusetts colonel recalled carrying a flag from Sarah "many days in Virginia and many miles up and down the valley of the Shenandoah."[2]

The relaxed social codes of the war allowed her to mingle among the officers and absorb their attention in a way that delighted her, and she enjoyed joking with them in a free manner that would have shocked Pittsfield before the war. "If you and your officers will favor us with your company," she wrote from Broadhall to a colonel camped nearby, "we will try and 'have a little dance tonight, boys.' I am longing, too, to have those camp glee songs begin; and as your time is now short here, I do hope you will all come tonight."[3]

A few of the younger officers were so grateful for her attention that they fancied themselves in love with her, and despite her poor health, she valiantly tried to sustain the illusion that she was still the most desirable woman in the Berkshires. Most of her flirtations were harmless, but at least one young captain—a Yale man named Edward Nettleton—reacted with great seriousness and wrote her long, suggestive letters. "Nothing," he urged her from his post at Camp Morewood in New Orleans, "can be enjoyed *heartily* without more than one heart to enjoy it." She tried—a little late in the game—to make light of his infatuation, but he was under her spell and insisted that his interest in her was serious. "No," he told her. "I do *not* write so long letters to everyone—I never knew you to complain of that before."[4]

Her proudest moment of the war came in August 1863, when she played host to one of the legendary heroes of the conflict, young William F. Bartlett, then a colonel in his twenties whose bravery in the face of intense fire had left him severely wounded and missing a leg. He was a Harvard junior when he entered the war in 1861, and his battlefield heroics had raised him quickly in the ranks. All of Pittsfield turned out to welcome his regiment when it returned to the Berkshires after a bloody campaign, and Melville's name was prominent among those helping to decorate the streets for the parade.

The colonel and his parents were Sarah's guests for a week, and she spared no expense to honor him and his men. Broadhall was transformed into a victory palace one evening, with a blue light shining in every window. Chinese lanterns were strung across the grounds and the sky was illuminated by fireworks. Melville was so impressed by the young colonel's stoic courage that he wrote an admiring poem about him after the welcoming festivities had ended. In "The College Colonel" he gave Sarah's guest the highest compliment he could offer, honoring him as a fellow castaway:

> He brings his regiment home—
> 　Not as they filed two years before,
> But a remnant half-tattered, and battered, and worn,
> Like castaway sailors, who—stunned
> 　　By the surf's loud roar,
> 　Their mates dragged back and seen no more—
> Again and again breast the surge,
> 　　And at last crawl, spent, to shore.

Sarah also wrote poems about the war. As the scholar Stanton Garner suggested long ago, one of her verses may have given Melville part of the title for his book of war poetry, *Battle-Pieces and Aspects of the War*. She could have shared her poem privately, but it was also published in the local paper under the title "The Rebellion." The first few lines contain the phrase that later reappeared in Melville's title:

Painfully the people wait
For the news by flying car,
Eager for the battle's fate
And the aspect of the war.

When the *Pittsfield Sun* published her work it was always anonymously or with a cryptic "S.," or in one unfortunate case with her full initials misprinted as S.V.M. (Depending on her mood and circumstances, her handwriting could vary considerably, and it is easy to imagine that the first letter of her middle name Anne could be misread as a *V.*) Topical themes didn't suit her as much as lyrical verse on her favorite subject of nature. One poem of this kind, written in 1862, did eventually end up in an anthology under "Mrs. S. A. Morewood," but without a title. This short work is her conventional but moving hymn to autumn as a serene but somber season full of memories of better days. It pleads for "future strength" to face the ordeal of "future loss," and it holds out the hope that underneath "autumn's blight" is a flickering gleam of spiritual rebirth for "bodies worn and wasted." It ends in a vision of "light, that's born of our decay, / Light, that ne'er shall waste away."[5]

Sadly, even as she threw herself so completely into the drama of the war, Sarah was wasting away. She lost weight, grew alarmingly pale, and often thought she was near death. But she refused to retreat from life and surrender to her illness, and time after time she struggled back from some dire moment of sickness to recover and go on for a little while longer. Doctors could do little to help, and Rowland grew so concerned that he even turned for guidance to Rev. John Todd, who blithely suggested that a few weeks in the remote woods at a hunting lodge in the Adirondacks would be a great tonic for Mrs. Morewood. Reluctantly, she gave the lodge a try and soon came home feeling no better for the trouble.

When her mother, Sarah Paradise Huyler, died in August 1862, she took it as a sign that her own time was finally coming to an end. After the funeral, she wrote movingly to her oldest child, William: "I have not been at all well since—and feel as if I lived in an unreal world. My mother was coming to visit me during this month and now I am never to see her again in this world. This is all that makes death so terrible—the eternal separation from friends."[6]

She loved the world and couldn't bear the thought of losing it. What helped her more than anything during the war was the joy of being the center of attention at large gatherings of soldiers. She kept entertaining and corresponding with them and giving them small presents until she had just weeks to live. Not even the smallest demands escaped her attention. Only two and a half months before her death the local paper reported, "Mrs. J.R. Morewood, who is never weary in the cause of well doing, has sent to the Soldiers' Relief Association a liberal donation of jellies, raspberry vinegar and sweetmeats."[7]

After struggling so long to keep her disease at bay, she faced death in 1863 at the height of autumn's beauty in the Berkshires. So great was her love for the region that, only a few days before she died, she gathered the strength for one last tour of the countryside. The early October weather was perfect, and she made a point of lingering over the views at her favorite hills. Riding in her carriage with her friend Caroline Whitmarsh, she covered eighty-five miles in a single outing and was so worn out by the end that she almost didn't make it home alive. "At each new turn of roads she knew so well," recalled Whitmarsh, "her sunken eyes would grow brilliant, and when too tired to speak, her languid hand pointed or grasped ours tightly; the little thin hand that had led so many toward happiness, and lifted so many from the dust. I feel it now!"[8]

Though it had been clear that she was fading fast, she had rallied so often that her death on October 16—one month and a day after her fortieth birthday—took everyone by surprise. Melville had left a few days earlier on a trip to New York, and was still there when the news came of her passing. Lizzie had visited Broadhall the day Sarah died and was at her bedside at the end. "She was much surprised herself when she knew her days were numbered—," Lizzie recalled afterward in a letter to Augusta. "Said she did not want to die, but was calm in view of it." On her last morning her sister opened a window in Sarah's room, and the dying woman raised her head to gaze one more time at the view. "How heavenly!" she said, then fell back and closed her eyes. Those were her last words. "Her respiration grew fainter & fainter," Lizzie remembered, "and so placid was her death, that no one knew exactly *when* she ceased to breathe."[9]

Devoted as always to his faith, Rowland accepted Sarah's death as God's will, and had sat peacefully near her while she was dying, busying himself with a letter to his son, William. Thinking at first that Sarah might survive the day, he began the letter by warning the boy—who was away at school in England—that the next letter he received would surely contain bad news. In Rowland's pious world, there was always a lesson to be learned from sad tidings. "Bear always in your mind," he wrote to his son, "that, though you are not likely to see her again on this earth, there is another world where you may again meet your Mother, if you rightly guide your own steps through life." Before he could finish this letter Rowland watched Sarah take her last breath. Instead of beginning a new message to William he put the news into a brief postscript: "3 o'clock P.M. . . . a change came over your dear Mamma's breathing, and she has now passed to the world of spirits."[10]

LIZZIE HAD NEVER seen anyone die before, and she said the experience left her shaken. She gave no hint in her letter to Augusta that she had any suspicions of an affair between Herman and Sarah, but she would spend the rest of her long life trying to pretend to the world that nothing was amiss in her marriage. At this point she may well have been fooled by all the precautions her husband had taken to hide his affair, and was still innocently assuming that his long attachment to Sarah was always just a friendship. That doesn't seem likely, however, and there is a forced note in Lizzie's declaration to Augusta that in her long relationship with Mrs. Morewood they never suffered "the least shadow of a break."

Only a private message that Lizzie carved into an inner compartment of her desk—crudely cutting the words into the wood with a sharp knife—gives some indication of the storms swirling below the surface of her married life. When she wrote the message isn't clear, but Melville's affair with Sarah continued to haunt him for the rest of his life. The desk is now at the Melville Room in the Berkshire Athenaeum, and the words in the dark recess of the compartment form a jagged line: "To know all is to forgive all." The sentence suggests that Lizzie may have always known more than she ever acknowledged, and that she struggled to make her peace with the facts.

Interestingly, even though servants came to prepare Sarah's body for burial, Lizzie insisted on remaining behind to help them. It is disturbing to think of her washing and dressing Sarah for the grave, especially in light of a jarring note of joy scrawled across one margin of Lizzie's otherwise dutifully solemn letter to Augusta. An old friend of the Shaws, the family doctor, had recently died and left Lizzie a generous bequest. Right next to her description of Sarah's final moments of life, Lizzie exulted in a chatty tone, "Did you know that Dr. Hayward left me a legacy of three thousand dollars? Nothing could have been more unexpected."[11]

No one but the two partners can know what really happens in a marriage, but Lizzie's letter on Sarah's death is eye-opening. As Somerset Maugham once noted, Melville's wife "was not a good letter-writer," but here she shows a disconcerting ability to switch her feelings on and off, and to allow personal pettiness to undercut a moment of deep sorrow and loss. Many people could overlook this sort of thing, but not Melville. He sought sympathy, warmth, and exalted feelings from Sarah, and perhaps he valued those from her all the more because he often failed to find them from Lizzie.[12]

ON THE DAY OF BURIAL, both Herman and his wife stood at Sarah's graveside in Pittsfield Cemetery. From New York, Melville had earlier sent a large floral wreath—all in white—for the church service he was forced to miss. In a newspaper tribute to Sarah a few days after her death, Caroline Whitmarsh predicted that the men whose lives her friend had "inspired" would not forget her. These included soldiers (men of "valor"), but rather daringly Whitmarsh wrote that Sarah had also inspired "men of genius." There weren't many of those in Pittsfield—except possibly a famous Harvard doctor who had moved away, and a novelist whose career had run aground in the fields of Arrowhead. If any friend knew how much Sarah had touched Melville's life, it would have been Whitmarsh. The wonder is that she was willing to hint at it in the pages of the Pittsfield press, and to add that Sarah's own talents for living large made an interesting contrast to the brilliance of more famous men—ones, for example, who wrote epics: "There is a genius that rears temples and writes epics; there is a better genius that makes all earth its temple, and all existence special. Such had Mrs. Morewood."[13]

Even if Whitmarsh wasn't also thinking of Dr. Oliver Wendell Holmes as one of the men of genius, he was indeed thinking of Sarah in the days after her death. Holmes couldn't forget his "Elsie Venner." He sent Rowland and the family a poem written in Sarah's memory. The little doctor was under her spell to the end, imagining himself looking down at her casket and finding nothing but a single rosebud.

> Oh, could it open into song,
>   How would its rosy heart-leaves tell
> That kindly thoughts are treasured long,
>   And loving deeds remembered well;
>
> And while the grace that Nature gives
>   Looks from the gentle downcast eye,
> The fruit, though perished, ever lives,
>   The flower, though faded, cannot die![14]

23

HOME FRONT

A few weeks after Sarah's death, Melville wrote to a young woman from Ohio who had visited Broadhall in recent summers and had attended some of the Morewood picnics at the family lake. She had sent him a request for a charity contribution and had addressed her letter to Pittsfield, but it had been forwarded to him at his new address on East Twenty-Sixth Street in New York. Informing her of the change, he wrote of his long stay in the Berkshires as if he had been there on an extended voyage—or, as he put it, a "visit." Twelve years was a long time to have been a Berkshire castaway: "Owing to my recent return to this, my native town, after a twelve years' visit in Berkshire, your note was delayed in reaching me."[1]

The move back to New York was only possible because of Lizzie. She had come into money—much more than she had received from the old family doctor. In 1861, at eighty, Lemuel Shaw had died, leaving her a good inheritance for the time—fifteen thousand dollars.

More important, not long before his death the judge had used all his considerable legal skill to disentangle Melville from debt and pay off what was left of the farm at Arrowhead. As a result, however, Melville was left with nothing. Judge Shaw managed to arrange things so that Lizzie—for her future well-being after he was gone—took possession of Arrowhead as her own property. By the time of Sarah's death, the place had been sold to Melville's brother Allan, and the move to New York was in the works.

So the Berkshires were home to Melville beginning with the Morewood purchase of Broadhall in 1850 and ending with Sarah's death in 1863. The last few years had been difficult because Sarah had so often been ill, and Herman was essentially living as a ward of his wife. He had earned nothing since 1860 and had no job in sight. He had to go where Lizzie led him, and she was enjoying her hold on the family budget. In her dry chronology of Melville's life that she prepared in old age, she writes of 1863 as though she were his banker, not his wife. "He moved into a house in New York—104 East 26th St bought from his brother Allan giving $7,750 and the Arrowhead estate valued at $3000 and assuming a mortgage of 2000 to Mrs. Thurston which was afterwards paid off by Dr Hayward's legacy to me of $3000 in May 1864—about $1000 [from] Aunt Priscilla's legacy was spent in repairs."[2]

After he left the Berkshires to live in his New York townhouse, Melville worked on his Civil War poems for *Battle-Pieces*, but for many months his heart wasn't in his work. As the reality of losing Sarah began to sink in, and the grimness of city life impressed itself on him, he didn't feel like doing anything. He was lonely, but that was now increasingly by choice. As the smoke cleared over the rubble of his career, he had tried in the later part of the 1850s to repair his relations with the Duyckincks, but there wasn't much they could do for him. Perhaps it was a sign of his own desperation that he had even made the effort to renew the friendship. It was gracious of him to try, and their past differences were quietly buried, though there was no pretending that things could ever be the same again. When Evert asked him to review a book shortly after Sarah's death, he declined, saying, "I have not spirit enough."[3]

THERE MUST HAVE BEEN MANY TIMES when Melville looked back on the storm that *Moby-Dick* churned up for him and wondered what it all meant. There can be no question that the sacrifice was enormous. The move back to New York in 1863 brought nothing but greater heartache, at least in the first several years. Once Sarah was no longer in his life, Melville saw his marriage take a decidedly acrimonious turn. There was no escape from the brutal truth of their basic incompatibility. Melville retreated into himself, and everyone suffered.

Still only in his forties, he was gripped by a simmering rage over the losses he had suffered. He didn't seem to care much for anything, including his own children by Lizzie. He had never been an attentive father, and now he was an angry one. As they grew older, his sons avoided him, and his daughters were wary of doing anything to provoke him. When his youngest child, Frances, was interviewed in the twentieth century, she was asked, "Did he rail at things in general when he was angry, or were his attacks more personal?" She responded with one word only. "Personal."[4]

Having nowhere to turn, he took out his frustrations on Lizzie. He went from being a chronically preoccupied, often distant husband to a bitter and utterly impossible one. At one stage of his decline in the 1860s, Lizzie's family begged her to leave Melville, and her brothers were eager to help in any way possible. Because the old judge was no longer in the picture, there was no one who seemed capable of reasoning with Melville.

A Unitarian minister in New York was asked to help arrange a separation if Lizzie agreed. One of her brothers—Samuel Shaw— told the minister in May 1867, "The thing has resolved itself into the mere question of my sisters willingness to say the word. . . . If you can suggest any plan of action by which the present lamentable state of things can be ended it will be most gratefully received." As her brother confided, the family was now firmly convinced that Melville had lost his mind, and that the only thing preventing Lizzie from leaving him was her anxiety over the way it would look "in the eyes of the world, of which she has a most exaggerated dread."[5]

Lizzie did seek help from the minister, Dr. Henry Bellows, but only so she could unburden her sorrows to him. A public separation from her husband at this late date was a step too far for this proud daughter of Judge Shaw, but she left no doubt in Dr. Bellows's mind that her marriage had become a "trial" to her. "And whatever further trial may be before me," she wrote to him, "I shall feel that your counsel is a strong help to sustain, more perhaps than any other earthly counsel could."[6]

WE CAN ONLY IMAGINE how painful the marriage had become to both husband and wife, and it is probably no coincidence that this turmoil became almost unbearable in the first few years after Melville lost Sarah. He must have felt at times that he had nothing left to live for. There was a hole in his life that no one could fill. On an allowance that Lizzie gave him, he was able to buy books and engravings, but he had no one with whom he could share them. One day in May 1867—at the height of his conflict with Lizzie—he acquired a book of poetry that affected him profoundly, largely because it brought his secret life with Sarah so vividly to mind. It was an old volume of verse translated from the sixteenth-century Portuguese poet Luis de Camões, who inspired Elizabeth Barrett Browning's *Sonnets from the Portuguese*.

In the introduction, Melville underlined a remark about Camões that shows how much, in his current distress, he was missing Sarah. "Woman was to him as a ministering angel, and for the little joy which he tasted in life, he was indebted to her." There was no joy with Lizzie, and like many couples who stay together when they should be apart, they were making each other miserable. With every argument, every unkind word, memories of the lost life at Broadhall must have heightened Melville's fury. It doesn't excuse it—it simply explains it.

The poems of Camões brought back all the old arguments for love that Melville had rehearsed in *Pierre*—the question of whether love is ever wrong, and the magnetic attraction of one lover to the other because of a glance or the sudden turn of a face. It was in Camões's "Madrigal" that the phrase "sweetest eyes" seemed to recall Sarah for Melville, who marked the poem in his copy:

> And sure if Love be in the right,
> (And was Love ever in the wrong?)
> To thee, my first and sole delight,
> That simple heart must now belong—
> Because thou hast the fairest mien
> And sweetest eyes that e'er were seen![7]

After reading accounts of the Portuguese poet's career, Melville decided to write his own poem about Camões because he thought their experiences were similar. Some of the facts were questionable, but he was drawn to the idea of Camões as a young seafaring adventurer who tasted love and fame and died with neither. Melville's words are supposed to describe the neglected writer in his unjust obscurity, but they are really a lament for his own fallen state. They are some of the most beautifully concise lines he ever wrote:

> Vain now thy ardor, vain thy fire,
> Delirium mere, unsound desire;
> Fate's knife hath ripped thy chorded lyre.

THE RIFT IN MELVILLE'S MARRIAGE may never have healed, but the turmoil did seem to abate after one of the children died. One day in the late summer of 1867, when the couple continued to be at odds, and the storm around them raged relentlessly, their older son, Malcolm, who was born in the heyday of Melville's literary success, shot himself in the bedroom of the family's New York house. He was only eighteen.

Melville was sobered by this tragedy. His son, he confessed too late, "never gave me a disrespectful word in his life, nor in any way ever failed in filialness." Gazing down on the boy as he lay dead, and seeing the look of peace on his son's face—"the ease of a gentle nature"—the father tried to find some comfort to soften the blow of his own failure as a parent. At first the death was ruled a suicide, but it was judged shortly afterward to be an accident—partly because of pressure from the family, all of whom wanted to deny that Malcolm had intended to kill himself.[8]

This was the lowest point in the marriage, and both husband and wife must have understood in the aftermath that they had to declare a truce. They couldn't change themselves or bring Malcolm back or find peace. They could only try to continue their lives without causing further damage. And so Melville retreated to his corner, Lizzie to hers, and they began the long march into an uneasy old age.

MELVILLE SURVIVED BOTH HIS SONS. The child born in the year that *Moby-Dick* was published—Stanwix—died of tuberculosis when he was thirty-four, far from home in San Francisco. But both of Melville's daughters would outlive their father. The older, Elizabeth, suffered from bad health most of her life, and never married. She died in 1908, when she was in her mid-fifties. Only one of Melville's children would enjoy a long life. Younger daughter Frances married, had four girls, and lived into her eighties. By the time she died in 1938, her father's name would finally be famous again, but she would say that this more celebrated figure was one she didn't recognize. "I don't know him in the new light," she remarked.[9]

24

LETTING GO

Melville did find employment to fill his later years, but it was nothing he was proud of. At the end of 1866 he accepted the burden of a new job that was an especially humiliating one for a writer of his talent, but he wanted the money. It was a position as a customs officer in New York, earning four dollars a day. It wasn't such a bad job, but what a fall it signified from the heights of his celebrity, and the high ambitions he had set for himself as a young man.

When he reported to his first post on the Hudson docks at the foot of a street with a name sure to haunt him with thoughts of lost Dutch-American glory—Gansevoort—one of his fellow officers recognized him, and was astounded to see him forced into taking the job. The man was a writer himself—Richard Henry Stoddard—and could remember meeting Melville many years earlier when the author was at the top of his career. "No American writer," he recalled, "was more widely known in the late forties and early fifties in his own country and in England than Melville." For almost the next twenty years the once-acclaimed author would labor in obscurity at his post.[1]

Business-minded Lizzie was pleased that her husband had finally been able to stick to a real job and see it through. Because of the money inherited from her family, she and her husband were finally able to live in reasonable comfort. In a letter to a relative on the occasion of Herman's retirement she wrote glowingly of his success as a customs man: "This month was a good turning-point, completing 19 years of faithful service, during which there has not been a single complaint against him—So he retires honorably of his own accord—He has a great deal [of] unfinished work at his desk which will give him occupation, which together with his love of books will prevent time from hanging heavy on his hands."[2]

Well meaning as these comments are, they give a good indication of why Melville so often struggled in his marriage. In her letter, the routine of toiling for nineteen years at the same job without "a single complaint" is a matter of proud achievement, whereas those books and the unfinished literary work in his study are primarily diversions to keep him from being bored. It brings to mind again that moving comment from Melville's *Clarel*: "My kin . . . would have me act some routine part. . . . This world clean fails me; still I yearn." If the world had not so completely failed him, his "honorable" career at the customs house would, in fact, have been filled with complaints. Angry ones at that, from any intelligent readers demanding to know why the author of *Moby-Dick* was "buried in a government office." But he didn't seem willing to fight the battle himself. For his family's sake, he accepted his fate, and came out on the other end of nineteen years like a prisoner ready to be released on good behavior.

IN RETIREMENT FROM HIS CUSTOMS POST, Melville could be forgiven for thinking his previous career as a professional author belonged to ancient history. He had written about the sea so long ago that he could be excused for thinking it belonged to some vanished dream of authorship. If asked about his days as a famous writer, he would shrug and pretend that it was all so long ago he couldn't remember much of it. "He seemed to hold his work in small esteem," recalled an old naval veteran, Peter Toft, who knew him in his last years, and who was one of his most enthusiastic champions. Melville would resist any attempt to discuss his books, telling his friend, "You know more about them than I do. I have forgotten them." But Peter Toft was not easily fooled by such talk. He was himself a survivor of the same maritime world that Melville had known in the 1840s, and he admired the author's major books precisely because they captured so vividly the life of that period. A few years after the author's death, Toft wrote in the *New York Times*, "Like Melville, I have also in my youth had a brief experience in a merchant ship, a Yankee whaler, and an American man-of-war. As a sailor boy in the maintop of the United States ship *Ohio* I was fascinated by his *Typee*, *Omoo*, *White-Jacket*, and his weird *Moby-Dick*."

With such a man, Melville couldn't resist discussing old times, but the books were another matter. The subject was too sensitive, not because he had forgotten them, but because the world had. "Melville, I understand, deliberately effaced himself in his latter years," said Toft, "and was naturally left severely alone, but I accidentally discovered him some years ago during my stay in New York, and, having much in common, we became good friends. Though a delightful talker when in the mood, he was abnormal, as most geniuses are, and had to be handled with care." One valuable insight that Toft managed to bring away from their talks was a strong sense of the affinity between the "weird romance" of *Moby-Dick* and the art of J. M. W. Turner. His new friend noted of this connection: "Melville, like Turner, delighted in 'color,' and sometimes in lurid color." At a time when Melville's work was relegated to the lower ranks of men who had written old-fashioned sea adventures, Toft was making an extraordinarily ambitious claim for his friend's talent, dropping his name alongside that of the great Turner. It would take at least another generation before anyone would entertain that connection seriously.[3]

BY THE END OF MELVILLE'S LIFE, the lingering taste of his earlier fame had turned acrid. Told of another writer who had yet to achieve fame, he scoffed, "What of that? He is not the less, but so much the more. . . . The further our civilization advances upon its present lines so much the cheaper sort of thing does 'fame' become, especially of the literary sort."[4] In his spare time Melville continued to write poetry, and to occasionally publish it in volumes that didn't sell. His verse epic, *Clarel: A Poem and Pilgrimage in the Holy Land,* occupied him for many years after his Mediterranean trip in 1856–57. It has moments of rare beauty, but as one contemporary reviewer pointed out when the poem appeared in 1876, the reader will have to climb over a "mound of sliding stones and gravel in the search for the crystals which here and there sparkle from the mass." The last of Melville's books published in his lifetime was a volume of poems in a private edition of twenty-five copies, with the unpromising title *Timoleon and Other Ventures in Minor Verse.*[5]

One reminder of the loss of Sarah may have found its way into the dense and confusing *Clarel* when its young hero contemplates at Easter the end of a thwarted romance and the death of the girl he loved, a woman named Ruth. In the wake of her loss everything seems hollow, especially at Easter in the Holy Land as she lies in her grave. She is sorely missed, yet even the cherished memory of her face is beginning to fade. The young man grieves that there is no hope for her return from the "prison" of the grave: "Christ is arisen; / But Ruth, may Ruth so burst the prison?"

AFTER MOVING TO NEW YORK, Melville only occasionally returned to the Berkshires. His association with the area began to disappear from the memories of the locals, except among a few old friends he had shared with Sarah, such as J. E. A. Smith, who always did his best to remind Pittsfield that the author of *Typee* once roamed among them. Broadhall remained in the Morewood family until around the end of the century. Rowland was a figure of some envy for Herman simply because the scenes of so many happy times still surrounded him at Broadhall. He never remarried, never seemed to take any interest in doing so, and was content to follow the usual twin passions of his life—business and religion.

When Melville sent his toast to Broadhall for the marriage of Anne Rachel, he also included a toast to Rowland. He made a point, not too subtly, of reminding him that Broadhall had once been his own paradise, and now what was left belonged to the widower. In the end Sarah was a bond that united these two very different men. "He ought to be a Happy Man," Melville's toast began, "for all he looks on without and within, is like a Paradise: and what is better, he *deserves* to be!"

So he did his duty in sending his best wishes to Rowland and Anne on her happy day, but he didn't think his job was complete unless he wrote a "Final & Concluding Toast." This one was meant for Sarah, whose spirit he guessed must surely be lingering somewhere in Broadhall that day. For a man who was never religious, but certainly spiritual, it was as close to a gesture of faith as he would ever make. Hoping that his words would reach her in the house that had meant so much to them, he tried summoning her spirit. It was a tribute to her enduring presence in his life. He filled it with the lavish praise he so often gave her. Someone was delegated to read his words to the assembled guests: "If there be a Spirit in this Company who seeks the pleasure of others before her own, whose delight is in happy faces about her, who forgets not friends far away, and whom no acquaintance with the world can make worldly or selfish—as one who is distant from the scene now clearly sees there is—at this moment of parting be she now remembered by us all as we drink, from the heart."[6]

Not only is the spirit a "she," but Melville also uses the present tense to make her seem alive, and then interrupts the listing of her virtues to affirm that even though he is "distant" from the wedding he "now clearly sees there is" indeed "a Spirit in this Company." He answers his own summons, offering the wedding guests a vision of Sarah whether they know to look for her or not.

IN OLD AGE Melville was exercising all the imagination he could bring forth to "clearly see" Sarah. At his death he left behind a poem about it. The setting is the wooded shore of their favorite lake north of Pittsfield, just below Greylock. At one point he thought of calling it simply "The Lake," but then—with his own eccentric spelling—he gave it the title "Pontoosuce," adding an *e* to the real name.

The poem takes place on a brilliant day at "autumnal noon-tide." The poet stands above the lake and admires its gleaming surface and the rich colors of the woods surrounding it. The more he thinks about the scene, the darker his thoughts become. The dying leaves give him an overwhelming sensation that death conquers everything, and that nothing survives. "The workman dies, and after him, the work," he says. Just when he is slipping into a moment of deep despair, convinced that "even truth itself decays," a vision comes to him of a beautiful woman whose form is bathed in a soft light like "the pale tints of morn." She is a dryad who emerges from a glade with a song on her lips, and she "floats" toward him with a wreath of pine sprigs to "her brow adorn." Her message, sung like a hymn, is the same as one of Sarah's later poems about autumn as a season that holds the promise of rebirth. ("Light, that's born of our decay," as Sarah expressed the idea in the work published as a hymn shortly after her death.) In Melville's poem the woman sings of nature, "Over and over, again and again, / It lives, it dies and it lives again."

To comfort the poet, she comes closer and hovers near his face to tell him in a soft voice not to shed tears for the dead. They have simply moved into another sphere of life. "All revolves," she says, "no more ye know." Then she whispers, "Let go, let go!" Coming even closer, she kisses him, and the "cold" wreath on her brow brushes against his, so that for a second the chaplet—as Melville calls it—seems to crown them both. Then she vanishes as abruptly as she appeared, leaving him feeling a combination of "warmth and chill," as if "life and death" had just been joined together—or, as he says, "wedded."[7]

It is possible to see Sarah in this poem, with its image of a wreath uniting two lovers, and to feel a sense on the poet's part that they should have been joined together in life and death. After Melville's career as an author of prose fiction fell apart and he turned to poetry, he rarely achieved the lyric grace and mystical power that he so movingly demonstrates in this poem about the woman who touched his life in ways that will never be fully understood. She was indeed his muse—his "goddess"—and the greatest love of his life, and hardly anyone knew it.

Sarah Morewood gave Melville a sympathetic ear just when he needed it most for the greatest challenge of his life—the writing of *Moby-Dick*. She was there to crown his success, regardless of what the world said, just as Hawthorne loyally wrote to him before leaving the Berkshires to say how much he admired *Moby-Dick*. Though the general reader will always think first of Melville's years at sea whenever the topic of *Moby-Dick* comes up, the book's true home is the Berkshires. Without Hawthorne, and without Greylock's majesty—the mountain as well as the woman—Melville might never have written what stands now beyond dispute as one of the greatest creations of any mind.

25

THE HANGING

The bearded old man returning home from Central Park with his young granddaughter didn't attract any notice as he walked down Fifth Avenue. He was about seventy, wore a plain blue suit, and moved at a steady pace, though with the help of a cane. Under a soft black hat his keen, watchful eyes studied the crowds who surged past him on the busy street, none of them aware that this quiet, unassuming grandfather was Herman Melville, the once-popular author.

Though people remembered *Typee,* no one knew much about *Moby-Dick.* It was a heavy old book from the lost age of sailing ships that told the story of harpooners in small boats sent to kill whales. Now coal and petroleum were fueling America's rise in the last decades of the nineteenth century, and *Moby-Dick* was largely forgotten. Even the little girl at his side, Eleanor—his first granddaughter, the child of Frances Melville—didn't think of him in those days as an author. It wasn't clear what he did each day in the dark study of the old house at 104 East Twenty-Sixth Street, with its big mahogany desk and sagging shelves of books. "His own room was a place of mystery and awe to me," she would recall. "There I never ventured unless invited by him." But she knew early on that his younger years had been full of strange and magical experiences, and that he had roamed far and wide.

On their return from these outings together he would pause in the front hall and stare at an engraving of the Bay of Naples. She never forgot the odd, almost dreamlike way that he would point at it with his cane and say, "See the little boats sailing hither and thither." Words like *thither* seemed almost exotic to her when he spoke them, but that was long before she knew that one of his books was called *Mardi: And a Voyage Thither*. She was spellbound whenever he put her on his lap and told her "wild tales of cannibals and tropic isles." When he was finished with his storytelling, she liked to tease him by pulling his beard, and he obliged, though she squeezed it hard, finding it "firm and wiry to the grasp."[1]

The old fires that used to rage in his heart had subsided. He seemed reasonably at peace, at least on the outside. Occasionally, a young admirer who had stumbled across one of his old books in a dusty shop would show up at East Twenty-Sixth Street to pay homage to the forgotten genius. When he retired from the government in 1886, one of the papers in New York remarked on the news: "The author is generally supposed to be dead. He has, indeed, been buried in a government office." A reporter paid him a call and found the old man to be "a genial, pleasant fellow, who, after all his wanderings, loves to stay at home."[2]

After retiring, Melville had one last chance to add another treasure to his prose works. At his desk in his study, he kept a quotation from Friedrich von Schiller: "Keep true to the dreams of thy youth." In his story of the life and death at sea of a simple young British sailor—the short work that would become *Billy Budd*—Melville set out to explore what it meant to leave behind in life a defining moment of greatness. In *Moby-Dick* it is the awful wreckage of Ahab's crazed voyage that takes the breath away. In *Billy Budd* it is the notion that some acts of sacrifice are so awe-inspiring that whatever the cost—however terrible the wreckage—they are worth it.

Billy goes bravely to his execution for killing a man on his ship who wronged him. Though his victim hated him and plotted to ruin him with false accusations, Billy's impulsive killing of the man is still murder and—according to the unyielding rules of the British Admiralty—the young man must be put to death. Accordingly, he is hanged from the mainyard, and afterward his body is sewn into his hammock and cast overboard. "The criminal paid the penalty of his crime," says a fictional naval publication that Melville cites in the aftermath of Billy's death. It is a severe, though correct, judgment.

Of course, the sailor is essentially an innocent man provoked into defending his honor. He never meant to kill his accuser when he impulsively lashed out and struck him. All the same, the law insists on sacrificing him to maintain its integrity, and the commander of Billy's ship—Captain Vere—will not do what his name suggests. He will not veer off course from the straight-and-narrow path of the law. It is his duty to hang Billy, and so he does. But whereas Captain Vere fails to rise to the occasion and sacrifice himself to save Billy, the young man proves with grace and dignity that he is better than the law that condemns him. The very simplicity of his character is his chief adornment as he embraces his sacrifice and dies like a hero instead of the criminal he is supposed to be in the eyes of the law.

As this late masterpiece proves, time wasn't "hanging heavy" on Melville's hands in retirement, as his wife sometimes feared. He was still wrestling with old questions and was still trying to come to terms with his own sacrifice on the altar of art. The world had condemned him for his supposed failures, and though his fate wasn't to die by a shot or at the end of a rope, he surely sacrificed his happiness and his family's happiness to write books that—as far as he could tell in the late 1880s—were mostly unread, and loved by only a few.

If he had learned anything from his career, it was that noble endeavors often suffer neglect and misunderstanding. Whatever whale he had been chasing in the most feverish and tumultuous part of his life, he had long ago lost track of it as it slipped beneath a wave and disappeared. The biggest question was always, Was it worth it? As an imperfect, indirect answer, *Billy Budd* offers a valiant yes. It is a powerfully imagined story that shows an old master back at work with undiminished powers. The manuscript is full of revisions and alterations, and he wouldn't have lavished so much care on it if he had not believed that his words still mattered. It was one last act of faith in a career that most people assumed was dead and buried. There were no more mad voyages to take, just one last story to tell in his best manner.

Uncomplaining and unafraid, Billy may die violently, but the moment of his execution is oddly quiet and peaceful. Instead of the all-consuming vortex swirling downward as it does in *Moby-Dick*, Melville offers a vision of Billy rising into a brilliant light. The rope that kills him also seems to deliver him into some greater realm. "Watched by the wedged mass of upturned faces, Billy ascended; and, ascending, took the full rose of the dawn." It is not, by any means, a vision of simple deliverance. The light may be there, high overhead, but Billy is still suspended by the neck, a sacrificial victim strangled like an animal. What is significant is that, at his death, he stares into an engulfing, frightening darkness as if it were light. That is his glory, his triumph over an uncomprehending world.

The fictional Billy Budd becomes a sacred hero to sailors. They even seek out relics. The spar from which Billy was hanged is so venerated among sailors that "a chip of it was as a piece of the Cross." He is no savior, but the sailors do admire his example, knowing that they might face a similar crisis one day, and that their resting place might be at the bottom of a deep ocean far from home. Vulnerable and superstitious, the men on those old sailing ships needed all the comfort they could find in their small keepsakes. Billy's relics were silent witnesses to courage.

THE MANUSCRIPT OF BILLY BUDD became something of a relic itself before its posthumous publication. When Melville died at home in New York on September 28, 1891—at the age of seventy-two—the story was among his papers, still needing further work but complete enough for publication. It was in such a confusing state that Lizzie wasn't sure how to deal with it, and at some point it was stored away in an old tin box. After Lizzie died in 1906, the manuscript moved around among family members until it came to rest in an attic of the New Jersey home belonging to his only married daughter, Frances. The box and its contents stayed in that attic for a decade, dust everywhere, and there was not much interest from the outside world in saving the literary remains of one Herman Melville.

Almost thirty years after he died, his granddaughter Eleanor rescued the forgotten manuscript that would help to bolster Melville's reputation in the twentieth century. Her mother allowed her to retrieve it from the attic, and to keep it in her possession. In 1919, when Eleanor was thirty-seven, a young scholar from Columbia University—Raymond Weaver—came to her looking for help with a biography he intended to write of the novelist, and she surprised him by revealing that her grandfather had left behind a substantial manuscript. As relics go, it was in good shape, but Weaver found it hard to determine the correct arrangement of the various parts. Though his edition of *Billy Budd* was far from perfect, its publication in 1924 helped to restore Melville's reputation, as did Weaver's biography of the writer, which came out in 1921.

Disillusioned, alienated, bitter—Melville was a genius just right for adoration from the readers of the next two centuries. His reputation has remained high ever since, his works having found a new generation of admirers in the 1920s. *Moby-Dick* was the great beneficiary of this revival. Survivors of the horrors of the First World War— the "Lost Generation"—didn't have any trouble understanding how Captain Ahab's dark battle against a monstrous foe could become an all-consuming descent into madness and destruction.

The Melville revival was a little slow getting off the ground in America, but British readers quickly rallied to the cause, and praise for the forgotten author poured from the London press. In 1927 the British novelist E. M. Forster was among the earliest critics to argue for the brilliance of *Billy Budd*. In his landmark work *Aspects of the Novel,* he treated the story as if the author's genius had always been apparent. "*Billy Budd* is a remote unearthly episode," wrote Forster, "but it is a song not without words, and should be read both for its own beauty and as an introduction to more difficult works. . . . Melville—after the initial roughness of his realism—reaches straight back into the universal, to a blackness and sadness so transcending our own that they are undistinguishable from glory." The glory Melville had sought was finally his. Now the laurel wreath came from one of Britain's most influential literary voices.[3]

American writers soon chimed in with their tributes to Melville, including praise from a young author of thirty who was at the very start of his career as a novelist. In July 1927 the *Chicago Tribune* asked William Faulkner if he could name a book he wished he had written. He chose *Moby-Dick* and went into flights of fancy about its "Greeklike" beauty, but what he enjoyed most was the evocative quality of Melville's whale. "There's magic in the very word, A White Whale," said Faulkner. "White is a grand word, like a crash of massed trumpets; and leviathan himself has a kind of placid blundering majesty in his name."[4]

Though born several years after Melville's death, Faulkner could hear the author's music loud and clear. Still resounding, the trumpets of *Moby-Dick* play on. But all the praise heaped on Melville in the last hundred years came so late that it wasn't easy to reconstruct the old story of a young author in love. It was much more convenient to assume that *Moby-Dick*'s origins were forever lost in forgotten yarns of the sea, or in some obscure rage suffered by a writer who would mysteriously decline into a kind of madness. But all the while, as decades came and went, the outlines of Melville's past in the Berkshires slowly began to take shape. The story of Sarah and Herman has finally emerged from the shadows, thanks to generations of scholars uncovering the detailed information on which this book has been built. Like *Moby-Dick* itself, the love story stubbornly refused to die.

CODA

On a cold Sunday morning in Elizabeth, New Jersey, an old man stood unsteadily beside a pew at the Christ Episcopal Church, trying to catch his breath. The service had started and all eyes were on the front of the church. Then, there was a loud moan, and every head turned to see the old man fall to the floor. He was carried to the vestibule and efforts were made to revive him. But he was gone, dead of a heart attack at eighty-two.

It had been more than half a century since the English-born John Rowland Morewood had married his young Dutch girl from New Jersey and settled her in the house of her dreams in the Berkshires. Now he lived alone, having retired and moved to Elizabeth, where he had some minor business interests. His fortune of earlier years was much diminished. "He was very wealthy at one time," the New York *World* remarked in its report of his death. There was no mention of his wife, who had been gone for forty years and was mostly forgotten. Two children were said to survive him, Mrs. Anne Lathers and Mr. William B. Morewood. A third child—Alfred—had died at home at the young age of thirty-two.[1]

They buried Rowland under a large white cross in the Pittsfield Cemetery, beside Alfred and Sarah Anne. The passage of time has left all the stones in bad shape, but most especially Sarah's, which has broken in half so that one part rests against the other, and the name and dates are hard to make out. Meanwhile, Melville and his wife are spending eternity in the Bronx, where they are buried at Woodlawn Cemetery. Messages from admirers often sit on the top of Herman's headstone, anchored by rocks.

Thanks to history enthusiasts in Pittsfield, Sarah Morewood is no longer an unfamiliar name to many local people. Everyone knows the country club, and the fact that the Morewoods used to own the whole spread on the south side of town is a matter of some curiosity. In 2006 the historical society gave twilight tours of the cemetery, and Sarah's crumbling grave was one of the stops. About seventy people showed up one summer night to see where—as the local paper put it—"coquettish socialite Morewood" was buried. And, strange to say, the highlight of one of these evenings was an appearance by Sarah herself. Not a ghost saying, "Let go," to a bearded stranger in the grip of an obsession. Just a local woman named Judy Daly who was wandering the grounds in costume. Someone had the inspiration to send out several actors dressed as departed figures from Pittsfield's history, each giving a little talk beside the grave of the famous person. How charming. Sarah would have loved it. One last costume party on a summer night in the Berkshires.[2]

ACKNOWLEDGMENTS

From the earliest stages of this book I have been fortunate to have the wise guidance and support of two remarkable women at the Friedrich Agency in New York, Molly Friedrich and Lucy Carson. This book could not have been written without their kind encouragement and advice.

I'm also grateful for invaluable editorial help from a great team of professionals at Ecco Books—Daniel Halpern, Hilary Redmon, and Emma Janaskie. I deeply appreciate their enthusiasm for this book, and their hard work on its behalf. Many thanks as well to Tom Pitoniak, for helpful suggestions when the book was still in typescript.

The marvelous collection of Melville documents and artifacts at the Berkshire Athenaeum in Pittsfield, Massachusetts, was essential to my book, and I feel especially fortunate to have worked there with Kathleen M. Reilly, supervisor of the Local History Department. She is a model of efficiency and a great supporter of Melville scholarship. On every visit to the Athenaeum I was encouraged by her warm welcome and generous assistance.

My time at Melville's old home of Arrowhead was memorable and enormously useful, thanks in large part to the helpful staff from the Berkshire County Historical Society—especially Will Garrison, curator, and Eileen Myers. I want also to thank Wayne Myers, for taking me to see Lake Morewood at what used to be the Broadhall farm. Members of the staff at the Pittsfield Country Club were also kind enough to show me the interior of the old mansion.

For research assistance, I am grateful to Michael Frost, Sterling Memorial Library, Yale University; Elizabeth E. Fuller, librarian, Rosenbach Museum and Library, Philadelphia; and Tal Nadan, reference archivist, Manuscripts and Archives Division, New York Public Library. Decades of scholarly research have been enormously helpful to me, especially the admirable labors of Harrison Hayford, Brian Higgins, Lynn Horth, Jay Leyda, Henry A. Murray, Steven Olsen-Smith, Hershel Parker, and Merton M. Sealts Jr.

I also appreciate the support of Gayle Cook; Professor Thomas Derrick; Mary Ann Duncan; Professor Joseph Fisher and Nancy Fisher; Maria McKinley and Dr. Lee McKinley; Professor Robert Perrin; Lee Pollock; Mary Burch Ratliff and Dr. Wesley Ratliff; June Shelden; Vanessa Shelden; and Sarah Shelden Coplen and Collins Coplen.

The dedication acknowledges imperfectly how much I owe to the love and influence of my wife and my grandparents.

NOTES

ABBREVIATIONS

BA: Berkshire Athenaeum, Pittsfield, Massachusetts.

BCHS: Berkshire County Historical Society, Pittsfield, Massachusetts.

HM: Herman Melville.

HMC: *Correspondence*. Ed. Lynn Horth. Evanston and Chicago: Northwestern University Press and the Newberry Library, 1989.

HMCR: *Herman Melville: The Contemporary Reviews*. Ed. Brian Higgins and Hershel Parker. Cambridge: Cambridge University Press, 1995.

HMJ: *Journals*. Ed. Howard C. Horsford and Lynn Horth. Evanston and Chicago: Northwestern University Press and the Newberry Library, 1989.

HMP: *Published Poems*. Ed. Robert C. Ryan, Harrison Hayford, Alma Mac-Dougall Reising, and G. Thomas Tanselle. Historical Note by Hershel Parker. Evanston and Chicago: Northwestern University Press and the Newberry Library, 2009.

HPBio: Parker, Hershel. *Herman Melville: A Biography, 1819–1851* and *Herman Melville: A Biography, 1851–1891*. 2 vols. Baltimore: Johns Hopkins University Press, 1996, 2002.

ML: Leyda, Jay. The *Melville Log: A Documentary Life of Herman Melville, 1819–1891*. New York: Harcourt Brace, 1951; New York: Gordian Press, 1969.

NYPL: Manuscripts and Archives, New York Public Library.

Rosenbach: Rosenbach Museum and Library, Philadelphia, Pennsylvania.

SAM: Sarah Anne Morewood.

Springfield: Lyman and Merrie Wood Museum of Springfield History, Springfield, Massachusetts.

Yale: Sterling Memorial Library, Yale University, New Haven, Connecticut.

=====

Because the major books Melville published in his lifetime are available in so many different editions, I cite only chapter numbers for these works in my notes and don't abbreviate the titles.

PROLOGUE

1. Maria Gansevoort Melville to Augusta Melville, December 29–30, 1851 (NYPL).
2. SAM to George Duyckinck, [December ?, 1851] (NYPL). Sarah Anne Morewood's middle name has occasionally appeared in print without the final *e,* but her gravestone in Pittsfield identifies her as "Sarah Anne," and she named her only daughter "Anne." A family record of birth dates lists her as the seventh of nine Huyler children who survived infancy, and gives her date of birth as September 15, 1823 (BCHS). The date is sometimes mistakenly given as 1824, which is the one on her gravestone.
3. See the November 1851 reviews reprinted in *HMCR* (386, 382, 378, 385, 380).
4. Maria Gansevoort Melville to Augusta Melville, December 29–30, 1851 (NYPL); G. Thomas Tanselle, "The Sales of Melville's Books," *Harvard Library Bulletin,* April 1969, 199.
5. The account of the Christmas dinner is taken from Maria Gansevoort Melville to Augusta Melville, December 29–30, 1851 (NYPL). This letter was first reported in 2002 in *HPBio,* 2:49. See chapter 18 here for a correction to the text. As a holy day of celebration, Christmas was gaining in popularity in New England, beginning at least in the early 1830s when Boston newspapers called for a "more marked observance of Christmas Day" (Restad, *Christmas in America,* 34). But celebrating the day was a long-established tradition among Episcopalians like John Rowland Morewood. As an example of SAM's floral talents, see her awards in "Report on Fancy Work, Drawings, Etc.," *Pittsfield Sun,* July 3, 1856.
6. Maria, HM's mother, refers to him as "very angry" in her letter to Augusta Melville of December 29–30, 1851 (NYPL), where she also describes Sarah's effort to crown HM: "She stopt before a plate on which lay a beautiful Laurel wreath, which she gently lifted & quickly placed upon his brow, he as quickly removed it to her head saying he would not be crowned."
7. See the index to Delbanco's *Melville,* 405. A fictional affair between Herman and Sarah is imagined in the middle section of Larry Duberstein's 1998 novel *The Handsome Sailor,* where a diary is invented to tell the

story. (See Jay Parini's review, "Call Me Herman," *New York Times*, June 28, 1998.) Hershel Parker is the only biographer to take a detailed look at the relationship between HM and SAM, but he dismisses any chance of a romance: "The wonder was that Sarah had not focused her attention on Herman in 1850. Nothing survives to indicate that she had done so" (*HPBio*, 2:44). Also, in the scholarly collection *Melville & Women*, Laurie Robertson-Lorant concludes, "As far as we know, Herman's relationship with Sarah Morewood was entirely platonic" (30).

8. Caroline Whitmarsh, "A Representative Woman," *Berkshire County Eagle*, October 29, 1863.

9. See HM to SAM, [December 20?,] 1853 (*HMC*, 252–55); and HM to SAM, September [12 or 19?,] 1851 (*HMC*, 205–6). The heavily annotated appearance of these letters in *HMC* may have led some scholarly readers to underestimate their romantic appeal, but seeing the handwritten words at BA left me with little doubt of HM's passionate exuberance. A recently recovered document at BA is a handwritten reply from "The Ladies of Arrowhead" to "Her Grace of Broadhall," accepting a dinner invitation in April 1852 from SAM. Though, like HM, the Arrowhead women (HM's wife, mother, and sisters) could imagine an aristocratic quality to SAM's reign at Broadhall, their response to "Her Grace" has none of the warmth or elaborate fancy that distinguish HM's letters to SAM.

10. HM to SAM, [August 29, 1856?] (*HMC*, 296–97). For SAM's family history see the Huyler files at BA. As an illustration of the courtly tradition of the laurel crown, see Anthony van Dyck's painting of Queen Henrietta Maria presenting a wreath to Charles I of England. Engravings and painted copies of this work were widely known in HM's time.

CHAPTER 1

1. Quoted in the Northwestern/Newberry edition of *Moby-Dick*, 611. For HM's memories of his uncle, see Merton M. Sealts Jr., "Thomas Melvill, Jr., in *The History of Pittsfield*," *Harvard Library Bulletin* 35 (1987): 201–17. (In the early 1830s HM, his mother, and siblings adopted the *e* at the end of the family name.)

2. *The WPA Guide to Massachusetts* cites Longfellow's term for what is now generally known as Morewood Lake.

3. Sealts, "Thomas Melvill, Jr., in *The History of Pittsfield*," 213, 214.

4. "Private Boarding House," *Pittsfield Sun*, August 9, 1849; Longfellow to Charles Sumner, July 23, 1848 (Longfellow, *Letters*, 3:177–78); Longfellow, *Life of Henry Wadsworth Longfellow*, 2:117–22.

5. SAM to her son William ("Willie") B. Morewood, [n.d.] (BCHS). Details of Sarah Huyler's early life are taken from documents at BCHS, including a handwritten "Family Record" preserved among the Morewood papers listing in order the birth dates of her parents and siblings.

6. SAM to George Duyckinck, [December 1851] (NYPL).

7. SAM to her sister-in-law Ann Morewood, October 24, 1848; and SAM to Susannah Perrin, [n.d.] (BCHS).

8. Seager, *And Tyler Too*, 319, 341.

9. Alexander Gardiner to David Gardiner, September 28, 1849 (Yale). See also *HPBio*, 2:42–43; and Parker, *Melville Biography*, 216. It isn't clear whether the compromising scene took place inside the mansion or elsewhere. Sarah was reportedly staying at another home for at least part of the 1849 visit.

10. SAM to George Duyckinck, July 29, [1856] (NYPL). The clergyman was Robert J. Parvin, who preached his last sermon in Pittsfield on July 20, 1856 (*Pittsfield Sun*, July 31, 1856).

11. Seager, *And Tyler Too*, 32.

CHAPTER 2

1. Published in 1851, the song "Fayaway" was written by Maria L. Child (lyrics) and J. F. Duggan (music). The quotation from *Typee* can be found in chapter 18.

2. *HMCR*, 44, 23; Wineapple, *Hawthorne*, 223.

3. HM to William B. Sprague, July 24, 1846 (*HMC*, 59).

4. *Typee*, chapters 32 and 2.

5. *HPBio*, 2:44, and *HMC*, 796; *HMCR*, 38–39.

6. "Our Ambrotypes. Herman Melville, Romanticist," *New York Daily News*, April 14, 1856, quoted in Steven Olsen-Smith, "Herman Melville's Planned Work on Remorse," *Nineteenth-Century Literature*, March 1996, 496.

7. *HMCR*, 157, 137, 130.

8. Quoted in *HMP*, 379; *Typee*, chapter 17.

9. *Typee*, chapter 18.

10. "Pacific Rovings," *Blackwood's*, June 1847 (*HMCR*, 120).

11. Henry Wadsworth Longfellow to Charles Sumner, July 23, 1848 (Longfellow, *Letters*, 3:177).

CHAPTER 3

1. James, *The American Scene*, 235.

2. Chase, *Lemuel Shaw*, 289, 282.

3. Ibid., 283.

4. Ibid., 294.

5. Fanny Appleton Longfellow, *Mrs. Longfellow*, 132.

6. *ML*, 259.

7. Quoted in Robertson-Lorant, *Melville*, 53.

8. *ML*, 59.

9. *HMCR*, 57.

10. Quoted in *HPBio*, 1:544.

11. Ibid., 1:554.

12. "After the Pleasure Party," *HMP*, 262; *ML*, 260; Parker, *Reading Billy Budd*, 45.

13. Wineapple, *Hawthorne*, 223. In a letter of May 31, 1818, Herman's father reported that his research in Edinburgh had uncovered the family's descent from Sir John Melvill of Carnbee, owner of Granton castle on the Firth of Forth (BA). The rock foundations of that castle are still there and overlook one of the great waterways of the world.

14. HM to Lemuel Shaw, October 6, 1849 (*HMC*, 138).

15. HM to R. H. Dana Jr., May 1, 1850 (*HMC*, 162).

16. "City Items: The Dog Days," *New York Tribune*, July 30, 1850.

CHAPTER 4

1. SAM to Susannah Perrin, [n.d.] (BCHS). For a typical description of the Morewood family's business in New York, see the advertisement "Rust Proof Iron," on the front page of the *New York Evening Post*, October 13, 1849. The Morewood family held valuable patents for galvanizing metal.

2. See Rowland's ad, "For Sale, A Modern Built Light Wagon," *Pittsfield Sun*, August 29, 1850.

3. In 1856 Dr. Oliver Wendell Holmes sold his house and 286 acres near Broadhall for $18,500 ("Farm Sold," *Pittsfield Sun*, August 21, 1856).

4. Caroline Whitmarsh, "A Representative Woman," *Berkshire County Eagle*, October 29, 1863.

5. No documents support the claim that Sarah's purchase of the mansion "stirred up in Herman a brew of feelings in which two of the ingredients were envy and jealousy" (*HPBio*, 1:735).

6. Elizabeth Shaw Melville to Hope Shaw, December 23, 1847 (quoted in *HMC*, 799).

7. "Several Days in Berkshire," *Literary World*, August 24, 1850; and Evert Duyckinck to Margaret Duyckinck, August 8, 1850 (see Luther Stearns Mansfield, "Glimpses of Herman Melville's Life in Pittsfield," *American Literature*, March 1937, 33).

8. For his description of Cornelius Mathews, see Lowell, *A Fable for Critics*, 36; HM, "Hawthorne and His Mosses."
9. HM to Evert Duyckinck, December 13, 1850 (*HMC*, 173); Parker, *Melville Biography*, 327.
10. "The Grand Fancy Dress-Ball," *Literary World*, September 7, 1850. Mathews's inscribed copy of *Chanticleer: A Thanksgiving Story* (1850) to "Mrs. Morewood" is held at BA.
11. *Pierre*, Book VII.
12. Ibid., Book V.
13. The Dryden edition is noted in Sealts, *Melville's Reading*, 57. The edition was published by Routledge in London in 1854. Also see the "Historical Note" in the Northwestern/Newberry edition of *Published Poems*, 416. The prominent markings in the margins of the Routledge Dryden are on pages 407 and 418. The rest are so faint that they almost escape detection by the naked eye, except for three words underlined on 440 ("mouths without hands") in "Cymon and Iphigenia." Next to these words, there is also a faint check mark in the margin. On page 415 in "Sigismonda and Guiscardo" a faintly marked passage includes the lines, "Permitted laurels grace the lawless brow, / The unworthy raised, the worthy cast below." All the prominent marks are consistent with those in other books owned by HM. (See the indispensable *Melville's Marginalia Online*, edited by Steven Olsen-Smith, Peter Norberg, and Dennis C. Marnon.)

CHAPTER 5

1. In 1850 HM was reading Smollett's *Roderick Random* in a copy borrowed from Evert Duyckinck (Sealts, *Melville's Reading*, 95). Smollett was already one of HM's favorite authors, and he had mentioned his work in three of the five books written before *Moby-Dick*: *Omoo* (see note 3 below); *Redburn* (chapter 29); and *White-Jacket* (chapters 8 and 12). For Scott on Smollett, see his *Lives of the Novelists*, 59. It has been suggested that SAM was thinking of Dr. Holmes's poem "Aunt Tabitha," but that would not have been possible in 1850 because the poem wasn't published until 1872. See Holmes, *Poetical Works*, 2:6–7, and Wyn Kelley, "Melville's Carnival Neighborhood," *Lectora*, 20 (2014). See the malapropisms of "our aunt Tabitha" in the Norton Critical Edition of *Humphry Clinker* (165, 279, 280, 12, 84, 280).
2. For identification of the costumes at the party, see *Melville in His Own Time*, 51.
3. Updike, *Hugging the Shore*, 93. Evan Gottlieb's Norton Critical Edition of *Humphry Clinker* spells Melville's surname with the final *e*, but the spelling

varies in earlier editions. For the reference to Smollett in *Omoo*, see chapter 77. Gottlieb's edition of *Humphry Clinker* reprints the Rowlandson engraving of the "venerable Turk" (120).

4. William Allen Butler to George Duyckinck, August 20, 1850 (see Luther Stearns Mansfield, "Glimpses of Herman Melville's Life in Pittsfield," *American Literature*, March 1937, 35).

5. *HMCR*, 398.

6. Mansfield, "Glimpses of Herman Melville's Life in Pittsfield," 35; Evert Duyckinck, "Notes of Excursions: Glimpses of Berkshire Scenery," *Literary World*, September 27, 1851.

7. Traubel, *With Walt Whitman in Camden*, 139.

8. Quoted in Mansfield, "Glimpses of Herman Melville's Life in Pittsfield," 28.

9. *Melville in His Own Time*, 49.

CHAPTER 6

1. Smith, *The Poet Among the Hills*, 151–54; Holmes, *Elsie Venner*, 56–57.

2. Holmes, *Elsie Venner*, 134, 137, 105–6. J. C. Burton, "Through the Berkshires," *Motor Age*, September 17, 1914; Smith, *The Poet Among the Hills*, 151–54. Though he makes no mention of a love affair between HM and SAM, see Rogin, *Subversive Genealogy* (184–85) for a more modern reference to the connection between SAM and *Elsie Venner*, and SAM and *Pierre*.

3. Holmes, *Elsie Venner*, 157, 57, 156.

4. Bacon, *Literary Pilgrimages*, 450. This book of 1902 also associates the Pittsfield area of South Mountain with the main setting of *Elsie Venner*.

5. Smith, *The Poet Among the Hills*, 151–54; and see John Dryden's translation of Virgil's *Aeneid*.

6. *ML*, 502.

7. "Forrest or Willis—Mr. Stevens's Card," *New York Tribune*, June 19, 1850.

8. Morse, *Life and Letters of Oliver Wendell Holmes*, 1:56; Holmes, *Elsie Venner*, 229, 113, 118, 115, 116, 131, 229, 132.

9. Hawthorne, *The American Notebooks*, 447–48.

10. *ML*, 636, and Gibian, *Oliver Wendell Holmes*, 3.

11. *Melville in His Own Time*, 172.

12. Holmes Jr. to Harold Laski, March 27, 1921 (*ML*, 936–37). In *ML* Jay Leyda suggests of Dr. Holmes and HM, "The two writers may have exchanged disguised portraits," in "I and My Chimney," and in *Elsie Venner* (xxviii).

CHAPTER 7

1. SAM to George Duyckinck, November 21, 1851 (NYPL). Emanuel Leutze painted Hawthorne's portrait in Washington, D.C., in 1862.

2. Wineapple, *Hawthorne*, 218.

3. *Melville in His Own Time*, 36.

4. Ibid., 75.

5. HM, "Hawthorne and His Mosses."

6. See HM's copy of *Mosses from an Old Manse* at melvillesmarginalia.org.

7. *Melville in His Own Time*, 74.

8. Ibid., 75.

9. Wineapple, *Hawthorne*, 226; Holmes, "At the Saturday Club," *Poetical Works*, 2:271.

CHAPTER 8

1. SAM to George Duyckinck, October 27, [1851] (NYPL).

2. "Dedication of the Pittsfield Rural Cemetery," *Pittsfield Sun*, September 12, 1850.

3. "Deaths," *Pittsfield Sun*, October 22, 1863.

4. Holmes, *Elsie Venner*, 197

5. "Dedication of the Pittsfield Rural Cemetery."

6. *HPBio*, 1:778; HM to Lemuel Shaw, May 22, 1856 (*HMC*, 295).

7. See the original newspaper passage, an excellent analysis of Smith's recollections, and early biographical work in Sealts, *The Early Lives of Melville*, 38–39, 130. In a letter to George Duyckinck (November 21, 1851), SAM had another explanation for "Broadhall" as the name of her house, saying that one of Melville's sisters had chosen the name (NYPL).

8. *HPBio*, 1:788, and *ML*, 396.

9. *Melville in His Own Time*, 76.

10. HM to Evert Duyckinck, October 6, 1850 (*HMC*, 170–71).

CHAPTER 9

1. HM to Evert Duyckinck, December 13, 1850 (*HMC*, 173–74).

2. Sealts, *The Early Lives of Melville*, 106.

3. Maria Gansevoort Melville to Augusta Melville, March 6, 1852 (NYPL).

4. Sealts, *The Early Lives of Melville*, 169.

5. HM to Evert Duyckinck, December 13, 1850 (*HMC*, 174), and *ML*, 404.

6. See "Knights and Squires," chapter 27 of *Moby-Dick*, for an extended treatment of these ideas.

7. Ibid., "The Quarter-Deck," chapter 36.

8. *Melville in His Own Time*, 30.

9. "Sunset," chapter 37, *Moby-Dick*.

10. HM to Evert Duyckinck, March 3, 1849 (*HMC*, 121).

CHAPTER 10

1. HM to Nathaniel Hawthorne, November [17?,] 1851 (*HMC*, 212).
2. Henry A. Murray, "In Nomine Diaboli," *Princeton University Library Chronicle*, Winter 1952, 47–62; Rebecca Stott, "*Moby-Dick*, into the Wonder-World, Audaciously," *You Must Read This*, with Robert Siegel, National Public Radio, June 13, 2007, and updated July 17, 2011.
3. HM to SAM, September [12 or 19?,] 1851 (*HMC*, 206).
4. The Melville copy of Todd's *Student's Manual* is held at BA.
5. Marvel, *Reveries of a Bachelor*, 67. See chapter 15 here for the evidence of SAM's awareness of Marvel's book.
6. HM to R. H. Dana Jr., May 1, 1850 (*HMC*, 162).

CHAPTER 11

1. Written in a note at the end of a poem whose first line is "The sky is clear. The moon her silvery light," SAM to George Duyckinck, [n.d.] (NYPL).
2. SAM to George Duyckinck, October 27, [1851] (NYPL).
3. HM to Evert Duyckinck, December 13, 1850 (*HMC*, 173).
4. "The Fountain," chapter 85, *Moby-Dick*.

CHAPTER 12

1. Review of HM's *White-Jacket* in the *Athenaeum*, February 2, 1850 (*HMCR*, 296). HM read of Turner's work as early as 1848 in Ruskin's *Modern Painters* (Sealts, *Melville's Reading*, 89).
2. Steven Olsen-Smith, "Melville's Copy of Thomas Beale's *The Natural History of the Sperm Whale* and the Composition of *Moby-Dick*," *Harvard Library Bulletin* 21 (Fall 2010): 1–77; "A Scamper Through the Exhibition of the Royal Academy," *Punch* 8 (1845): 233; and "A Peep into the Royal Academy," *New Monthly Magazine*, June 1845.
3. Leslie, *Autobiographical Recollections*, 138.
4. "Loomings," chapter 1, *Moby-Dick*.
5. Ibid., "The Pulpit," chapter 8.
6. Ibid., "The Funeral," chapter 69.
7. Quoted in Waid, *Edith Wharton's Letters from the Underworld*, 171.
8. "The Needle," chapter 124, and "The Chase—Third Day," chapter 135, *Moby-Dick*. For discussion of another painting in *Moby-Dick* with suggestions of Turner's style, see Robert K. Wallace, "Melville and the Visual Arts," in *A Companion to Herman Melville*, where Wallace writes about the "squitchy picture" of "unimaginable sublimity" in the Spouter-Inn (349).

9. Finley, *Angel in the Sun*, 179; Bell, *A List of the Works Contributed to Public Exhibitions by J. M. W. Turner, R.A.*, 155; Rogers, *The Voyage of Columbus* in *Poems*, 131.

10. HM's comment on Italy is quoted in *HMJ*, 368.

11. Roberts, *Samuel Rogers and His Circle*, 48; Cunningham, *Modern London*, 26.

CHAPTER 13

1. Julian Hawthorne, *Nathaniel Hawthorne and His Wife*, 1:376.

2. HM to Evert Duyckinck, February 12, 1851 (*HMC*, 180).

3. HM to Nathaniel Hawthorne, [January 29?,] 1851 (*HMC*, 176).

4. *Melville in His Own Time*, 147.

5. HM, *White-Jacket*, chapter 4.

6. HM to Nathaniel Hawthorne, [June 1?,] 1851 (*HMC*, 191–92); HM to Nathaniel Hawthorne, June 29, 1851 (*HMC*, 196).

7. *Melville in His Own Time*, 77. This is the most reliable version of Sophia Hawthorne's remarks.

8. Wineapple, *Hawthorne*, 345.

9. Hawthorne, *The House of the Seven Gables*, chapter 11.

10. HM to Nathaniel Hawthorne, [April 16?,] 1851 (*HMC*, 185).

11. Ibid. (*HMC*, 187).

12. HM to Nathaniel Hawthorne, [May ?,] 1851 (*HMC*, 191). This letter has been published with an indication that it was written in June, but May seems more likely (see *HPBio*, 1:840–41); and June 29, 1851 (*HMC*, 196).

13. HM to Nathaniel Hawthorne, [May ?,] 1851 (*HMC*, 191). For the dating of this letter, see the note above.

CHAPTER 14

1. HM to Evert Duyckinck, March 26, 1851 (*HMC*, 183). SAM's date of return from England can be determined from shipping news in various periodicals and in documents in BCHS, especially George Morewood to Rowland and Sarah Morewood, May 23, 1851.

2. HM to Nathaniel Hawthorne, June 29, 1851 (*HMC*, 195). See also *HPBio*, 1:839–40.

3. SAM to Susannah Perrin, [n.d.] (BCHS).

4. See *HPBio*, 1:852.

5. *ML*, 419.

6. HM to Nathaniel Hawthorne, July 22, 1851 (*HMC*, 199).

7. Smith, *Taghconic*, 156. As evidence of SAM's authorship of the Greylock essay, see *ML*, 461. Evert Duyckinck to Margaret Duyckinck, August 7,

1851 (Luther Stearns Mansfield, "Glimpses of Herman Melville's Life in Pittsfield," *American Literature*, March 1937, 39).

8. SAM to George Duyckinck, October 27, [1851] (NYPL), and *HPBio*, 2:46. For information about Rev. Entler, see his obituary in the *Congregational Year-Book* of 1887.

9. Evert Duyckinck to Margaret Duyckinck, August 13, 1851 (Mansfield, "Glimpses of Herman Melville's Life in Pittsfield," 44–45).

10. Holmes, *Elsie Venner*, 193.

CHAPTER 15

1. Thoreau, *A Week on the Concord and Merrimack Rivers*, 153.
2. *Melville in His Own Time*, 64; Smith, *Taghconic*, 155.
3. Smith, *Taghconic*, 152.
4. Ibid., 154.
5. Ibid., 156.
6. "The Candles," chapter 119, *Moby-Dick*.
7. For the use of the "motto" in *Moby-Dick*, see "The Forge," chapter 113.
8. Smith, *Taghconic*, 156. SAM to George Duyckinck, October 27, [1851] (NYPL).
9. *HPBio*, 2:534, and SAM to George Duyckinck, [December 24, 1851] (NYPL).
10. George Duyckinck to SAM, August 18, 1851 (BA); SAM to George Duyckinck, [December 22?, 1851], and August 24, [1853?] (NYPL).
11. Holmes, *Elsie Venner*, 293, 311.
12. *Obituary Notice of the Late George L. Duyckinck*, 17.
13. SAM to George Duyckinck, September 14, [1851] (NYPL).
14. SAM to George Duyckinck, December 28, 1851 (NYPL). For Evert's remark about "a collection of minerals," see "Marks and Remarks," *Literary World*, December 27, 1851 (508).
15. SAM to George Duyckinck, [December ?, 1851] (NYPL).

CHAPTER 16

1. SAM to George Duyckinck, [October 8, 1851] (NYPL).
2. Smith, *Taghconic*, 42.
3. In his *Melville's Prisoners* (132–83), the excellent scholar Harrison Hayford demonstrated conclusively why the various theories of Melville's "secret" sister won't stand up to scrutiny. For Hawthorne's attack on the Shaker men, see Hawthorne, *The American Notebooks*, 465.
4. *Pierre*, Book VIII.

5. Ibid., Books XXII and XXI.

6. Ibid., Book V.

7. Ibid., Book VIII. See Murray's "Introduction" to the 1949 Hendricks House edition of *Pierre* (liii).

8. *Pierre,* Books XII and II.

9. Ibid., Book VIII.

10. Ibid., Book XXIII.

CHAPTER 17

1. HM to Nathaniel Hawthorne, November [17?,] 1851 (*HMC,* 212).

2. Ibid. (213).

3. HM to Sophia Van Matre, December 10, 1863 (*HMC,* 387).

4. Mellow, *Nathaniel Hawthorne in His Times,* 376.

5. *HMCR,* 384, 415, 387, 397.

6. Ibid., 412, 378, 382, 380.

7. Ibid., 384–86.

8. Mellow, *Nathaniel Hawthorne in His Times,* 382.

9. *Obituary Notice of the Late George L. Duyckinck,* 9.

10. Richard Bentley to HM, May 5, 1852 (*HMC,* 620), and HM to Richard Bentley, July 20, 1849 (*HMC,* 133).

11. *HMCR,* 410.

12. *Pierre,* Book XXV.

CHAPTER 18

1. Lawrence, *Selected Literary Criticism,* 374.

2. For details of Stanwix's birth record, see *ML,* 430, and Hayford, *Melville's Prisoners,* 78–79. Hayford's book includes a helpful letter from the Pittsfield city clerk.

3. Lawrence, *Selected Literary Criticism,* 374. Maria's complaint against HM is quoted in *HMC,* 784.

4. Maria Gansevoort Melville to Augusta Melville, December 29–30, 1851 (NYPL).

5. HM to SAM, September [12 or 19?,] 1851 (*HMC,* 205–6). For the phrase "voucher of paradise," see HM's poem "The Devotion of the Flowers to Their Lady."

6. SAM to George Duyckinck, [January ?, 1852] (NYPL).

7. Maria Gansevoort Melville to Augusta Melville, December 29–30, 1851 (NYPL). In *HPBio,* 2:50, this passage in Maria's letter ends with her remark to SAM "that her conversation affected her." But, in fact, there is a

dash after "her," and the sentence continues at the top of the letter's first page, so that it reads "her conversation affected her Husband very painfully & I wished her to change the subject."

8. SAM to George Duyckinck, [December 28, 1851] (NYPL).

9. HM to the Editors of the *Literary World*, February 14, 1852 (222).

10. Matthiessen, *American Renaissance*, 471. SAM went to New York the first week of September, but returned to Broadhall September 11. See SAM to George Duyckinck, September 14, [1851] (NYPL), misdated 1850.

11. SAM to George Duyckinck, [September 27, 1851] (NYPL).

12. Wineapple, *Hawthorne*, 221.

CHAPTER 19

1. *HMCR*, 441.

2. Ibid., 433, 437, 424.

3. Ibid., 436, 433.

4. Ibid., 419–20, 421, 426.

5. *Pierre*, Book XXV.

6. Ibid., Book XXII.

7. Ibid.

8. Ibid., Book XXV.

9. Ibid., Books XVII and XXVI.

10. *HMCR*, 430–31.

11. SAM to George Duyckinck, December 28, 1851 (NYPL).

12. Maugham, *W. Somerset Maugham Selects the World's Ten Greatest Novels*, 218–19.

13. Smith, *Taghconic*, 151.

14. *Pierre*, Book XXV.

15. Sealts, *Melville's Reading*, 76.

CHAPTER 20

1. *ML*, 478–89.

2. Meyers, *Edgar Allan Poe*, 206.

3. *HMC*, 855. For corroboration, see note 6 in chapter 24.

4. HM to SAM, [December 20?,] 1853 (*HMC*, 252–53).

5. "Germany," *Christian Times*, January 2, 1852; Wilson, *An Encyclopedia of Continental Women Writers*, 1:523; Chambers, *Humor and Irony in Nineteenth-Century German Women's Writing*, 52; and "Foreign Items," *New York Tribune*, April 2, 1851.

6. *ML*, 469.

7. HM to SAM, December 2, 1860 (*HMC*, 357).
8. *Pierre,* Book XII.

CHAPTER 21

1. *ML,* 521 and 525.
2. Spark, *Hunting Captain Ahab,* 212, and Elizabeth Renker, "Herman Melville, Wife Beating, and the Written Page," *American Literature,* March 1994; HM to SAM, [August 29, 1856?] (*HMC*, 296–97).
3. *HMJ,* 628.
4. Ibid., 628–29.
5. HM to Sophia Hawthorne, January 8, 1852 (*HMC*, 219).
6. *HMJ,* 633.
7. Ibid., 129.
8. *ML,* 604–6.

CHAPTER 22

1. Caroline Whitmarsh, "A Representative Woman," *Berkshire County Eagle,* October 29, 1863.
2. Letter from Colonel A. Potter, *Pittsfield Sun,* February 2, 1871.
3. SAM to Lieutenant Colonel Wheldon, January 23, 1862 (Springfield).
4. Edward P. Nettleton to SAM, April 8, 1863 (BA).
5. "The Rebellion," *Pittsfield Sun,* January 15, 1863. Most of the unattributed poems in the paper can be traced to previously published sources, but of the few that are identified as "for the Sun" and carry no author's name in full, three or four seem likely the work of Sarah, whose contributions to the paper were warmly praised at her death. See Garner, *The Civil War World of Herman Melville,* 215, and Sarah's untitled lyric in *Hymns of the Ages,* Third Series (Boston: Ticknor & Fields, 1865), 261.
6. SAM to William B. Morewood, [September 1862] (BCHS).
7. August 6, 1863, *Pittsfield Sun.*
8. Whitmarsh, "A Representative Woman."
9. Elizabeth Shaw Melville to Augusta Melville, October 16, 1863 (NYPL).
10. J. Rowland Morewood to William B. Morewood, October 16, 1863 (BA).
11. Elizabeth Shaw Melville to Augusta Melville, October 16, 1863 (NYPL).
12. Maugham, *W. Somerset Maugham Selects the World's Ten Greatest Novels,* 218.
13. Whitmarsh, "A Representative Woman."
14. Holmes, "Written for S. A. Morewood" (BA).

CHAPTER 23

1. HM to Sophia Van Matre, December 10, 1863 (*HMC*, 386).
2. Sealts, *The Early Lives of Melville*, 174.
3. HM to Evert Duyckinck, December 31, 1863 (*HMC*, 389).
4. Spark, *Hunting Captain Ahab*, 214.
5. Samuel S. Shaw to Henry W. Bellows, May 6, 1867 (*HMC*, 858–59).
6. Elizabeth Shaw Melville to Henry W. Bellows, May 20, 1867 (*HMC*, 860).
7. See Monteiro, *The Presence of Camões*, 62–79.
8. Quoted in *HMC*, 400.
9. Elizabeth Renker, "Herman Melville, Wife Beating, and the Written Page," *American Literature*, March 1994. As Renker has correctly pointed out, "Critics have tried hard to redeem the Melville marriage as a good one by reading Melville's rose poems [late verses in *Weeds and Wildings*] as love poems to Lizzie, but neither the rose poems nor the biographical facts of the marriage sustain such a case" (*A Companion to Herman Melville*, 492). At any rate, HM associated the rose with SAM ("Most considerate of all the delicate roses that diffuse their blessed perfume among men, is Mrs. Morewood").

CHAPTER 24

1. Stoddard, *Recollections*, 143.
2. Elizabeth Shaw Melville to Catherine Lansing, January 10, 1886 (*ML*, 796).
3. Peter Toth, "In Praise of Herman Melville," *New York Times*, March 17, 1900.
4. HM to James Billson, December 20, 1885 (*HMC*, 492).
5. *New York Tribune*, June 16, 1876 (*HMCR*, 532).
6. *HMC*, 855–56. The toast to SAM's "spirit" ends with an awkward effort to return the toast to Rowland ("the Health of Mr. Morewood"). Because SAM is dead ("be she now remembered by us all"), this occasion can't be confused with another in 1851 when Cornelius Mathews sent the Morewoods a witty toast.
7. See Robert Penn Warren's edition of the poem in his *Selected Poems of Herman Melville*, 344–47. "Skies are dark and winds are moaning, / Leaves around us falling fast" are the opening lines of Sarah's untitled autumnal lyric in *Hymns of the Ages*, Third Series (Boston: Ticknor & Fields, 1865), 261.

CHAPTER 25

1. Weaver, *Herman Melville*, 377–79.
2. *HMC*, 728.

3. Forster, *Aspects of the Novel*, 142.

4. Faulkner, *Essays, Speeches & Public Letters*, 197–98.

CODA

1. "Dropped Dead in Church," *New York World*, January 26, 1903.

2. "Guests, Not Ghouls, on Grave Trip," *Berkshire Eagle*, September 9, 2006.

BIBLIOGRAPHY

Allen, Gay Wilson. *Melville and His World*. London: Thames & Hudson, 1971.

Arvin, Newton. *Herman Melville*. New York: William Sloane, 1950.

Bacon, Edwin M. *Literary Pilgrimages in New England*. New York: Silver Burdett, 1902.

Bailey, Anthony. *Standing in the Sun: A Life of J. M. W. Turner*. London: Tate, 2013 [1997].

Bell, C. F. *A List of the Works Contributed to Public Exhibitions by J. M. W. Turner, R.A.* London: George Bell, 1901.

Borst, Raymond K. *The Thoreau Log: A Documentary Life of Henry David Thoreau, 1817–1862*. New York: G. K. Hall, 1992.

Chambers, Helen. *Humor and Irony in Nineteenth-Century German Women's Writing: Studies in Prose Fiction, 1840–1900*. Rochester, NY: Camden House, 2007.

Chase, Frederic Hathaway. *Lemuel Shaw: Chief Justice of the Supreme Judicial Court of Massachusetts, 1830–1860*. Boston and New York: Houghton Mifflin, 1918.

Clayden, P. W. *The Early Life of Samuel Rogers*. London: Smith, Elder, 1887.

A Companion to Herman Melville. Ed. Wyn Kelley. Malden, MA: Blackwell, 2006.

A Companion to Melville Studies. Ed. John Bryant. New York: Greenwood, 1986.

Critical Essays on Herman Melville's Pierre; or, The Ambiguities. Ed. Brian Higgins and Hershel Parker. Boston: G. K. Hall, 1983.

Cunningham, Peter. *Modern London; Or, London as It Is*. London: John Murray, 1851.

Delbanco, Andrew. *Melville: His World and Work*. New York: Vintage Books, 2006.

Dolin, Eric Jay. *Leviathan: The History of Whaling in America*. New York: Norton, 2007.

Faulkner, William. *Essays, Speeches & Public Letters*. Ed. James B. Meriwether. New York: Modern Library, 2004.

Finley, Gerald. *Angel in the Sun: Turner's Vision of History*. Montreal: McGill-Queen's University Press, 1999.

Forster, E. M. *Aspects of the Novel*. New York: Harcourt, 1955 [1927].

Garner, Stanton. *The Civil War World of Herman Melville*. Lawrence: University Press of Kansas, 1993.

Gibian, Peter. *Oliver Wendell Holmes and the Culture of Conversation*. Cambridge: Cambridge University Press, 2001.

Habegger, Alfred. *My Wars Are Laid Away in Books: The Life of Emily Dickinson*. New York: Modern Library, 2002.

Hardwick, Elizabeth. *Herman Melville*. New York: Viking Penguin, 2000.

Hawthorne, Julian. *Nathaniel Hawthorne and His Wife: A Biography*. 2 vols. Boston: James R. Osgood, 1885.

Hawthorne, Nathaniel. *The American Notebooks*. Centenary Edition. Ed. Claude M. Simpson. Columbus: Ohio State University Press, 1972.

Hayford, Harrison. *Melville's Prisoners*. Evanston, IL: Northwestern University Press, 2003.

Holmes, Oliver Wendell. *Elsie Venner: A Romance of Destiny*. New York: Grosset & Dunlap, n.d. [1861].

————. *The Poetical Works of Oliver Wendell Holmes*. Boston: Houghton Mifflin, 1892.

Irmscher, Christoph. *Longfellow Redux*. Urbana and Chicago: University of Illinois Press, 2006.

James, Henry. *The American Scene*. New York: Harper & Brothers, 1907.

Lawrence, D. H. *Selected Literary Criticism*. Ed. Anthony Beal. New York: Viking, 1966.

Leslie, Charles Robert. *Autobiographical Recollections*. Ed. Tom Taylor. Boston: Ticknor & Fields, 1860.

Longfellow, Fanny Appleton. *Mrs. Longfellow: Selected Letters and Journals of Fanny Appleton Longfellow*. Ed. Edward Wagenknecht. New York: Longmans, Green, 1956.

Longfellow, Henry Wadsworth. *The Letters of Henry Wadsworth Longfellow*. Ed. Andrew Hilen. 6 vols. Cambridge, MA: Harvard University Press, 1966–82.

————. *Life of Henry Wadsworth Longfellow*. Ed. Samuel Longfellow. 3 vols. Boston: Ticknor, 1886–87.

Lowell, James Russell. *A Fable for Critics*. Boston: Houghton Mifflin, 1891 [1848].

Marshall, Megan. *Margaret Fuller: A New American Life*. Boston: Houghton Mifflin Harcourt, 2013.

Marvel, Ik [Donald G. Mitchell]. *Reveries of a Bachelor: or A Book of the Heart*. New York: Scribner, 1853.

Matthiessen, F. O. *American Renaissance: Art and Expression in the Age of Emerson and Whitman*. New York: Oxford University Press, 1941.

Maugham, W. Somerset. *W. Somerset Maugham Selects the World's Ten Greatest Novels*. New York: Fawcett, 1959 [1948].

Mellow, James R. *Nathaniel Hawthorne in His Times*. Boston: Houghton Mifflin, 1980.

Melville, Herman. *Selected Poems of Herman Melville: A Reader's Edition*. Ed. Robert Penn Warren. Boston: David R. Godine, 2004 [1970].

Melville & Women. Ed. Elizabeth Schultz and Haskell Springer. Kent, OH: Kent State University Press, 2006.

Melville in His Own Time: A Biographical Chronicle of His Life, Drawn from Recollections, Interviews, and Memoirs by Family, Friends, and Associates. Ed. Steven Olsen-Smith. Iowa City: University of Iowa Press, 2015.

Meyers, Jeffrey. *Edgar Allan Poe: His Life and Legacy*. New York: Cooper Square Press, 2000.

Monteiro, George. *The Presence of Camões: Influences on the Literature of England, America & Southern Africa*. Lexington: University Press of Kentucky, 1996.

Morse, John T. *Life and Letters of Oliver Wendell Holmes*. 2 vols. Boston: Houghton Mifflin, 1896.

Obituary Notice of the Late George L. Duyckinck, Esq. New York: Church Book Society, 1863.

Parker, Hershel. *Melville Biography: An Inside Narrative*. Evanston, IL: Northwestern University Press, 2012.

———. *Reading Billy Budd*. Evanston, IL: Northwestern University Press, 1990.

Paston, George. *At John Murray's: Records of a Literary Circle, 1843–1892*. London: John Murray, 1932.

Philbrick, Nathaniel. *Why Read Moby-Dick?* New York: Penguin Books, 2013 [2011].

Restad, Penne L. *Christmas in America: A History*. Oxford: Oxford University Press, 1995.

Reynolds, David S. *Beneath the American Renaissance: The Subversive Imagination in the Age of Emerson and Melville*. Cambridge, MA: Harvard University Press, 1989.

Roberts, R. Ellis. *Samuel Rogers and His Circle*. New York: Dutton, 1910.

Robertson-Lorant, Laurie. *Melville: A Biography*. New York: Clarkson Potter, 1996.

Robinson, Forrest G. *Love's Story Told: A Life of Henry A. Murray*. Cambridge, MA: Harvard University Press, 1992.

Rogers, Samuel. *Poems*. London: Routledge, 1890.

———. *Reminiscences and Table-Talk of Samuel Rogers: Banker, Poet, & Patron of the Arts, 1763–1855*. Ed. G. H. Powell. London: R. Brimley Johnson, 1903.

Rogin, Michael Paul. *Subversive Genealogy: The Politics and Art of Herman Melville*. New York: Knopf, 1983.

Ruskin, John. *Notes on the Turner Gallery at Marlborough House, 1856*. London: Smith, Elder, 1857.

Savage Eye: Melville and the Visual Arts. Ed. Christopher Sten. Kent, OH: Kent State University Press, 1991.

Scott, Walter. *Lives of the Novelists*. London: Oxford University Press, 1906.

Seager, Robert. *And Tyler Too: A Biography of John and Julia Gardiner Tyler*. New York: McGraw-Hill, 1963.

Sealts, Merton M., Jr. *The Early Lives of Melville: Nineteenth-Century Biographical Sketches and Their Authors*. Madison: University of Wisconsin Press, 1974.

———. *Melville's Reading: A Check-list of Books Owned and Borrowed*. Madison: University of Wisconsin Press, 1966.

———. *Pursuing Melville, 1940–1980: Chapters and Essays*. Madison: University of Wisconsin Press, 1982.

Smith, J. E. A. ["Godfrey Greylock"]. *The Poet Among the Hills: Oliver Wendell Holmes in Berkshire*. Pittsfield, MA: George Blatchford, 1895.

———. *Taghconic; or Letters and Legends About Our Summer Home*. Boston: Redding, 1852.

Smollett, Tobias. *The Expedition of Humphry Clinker*. Norton Critical Edition. 2nd ed. Ed. Evan Gottlieb. New York: Norton, 2015 [1771].

Spark, Clare L. *Hunting Captain Ahab: Psychological Warfare and the Melville Revival*. Kent, OH: Kent State University Press, 2001.

Stoddard, Richard Henry. *Recollections: Personal and Literary*. New York: A. S. Barnes, 1903.

Thoreau, Henry David. *A Week on the Concord and Merrimack Rivers*. New York: Library of America, 1985 [1849].

Traubel, Horace. *With Walt Whitman in Camden (March 28–July 14, 1888)*. New York: Appleton, 1908.

Updike, John. *Hugging the Shore: Essays and Criticism*. New York: Knopf, 1983.

Waid, Candace. *Edith Wharton's Letters from the Underworld: Fictions of Women and Writing*. Chapel Hill: University of North Carolina Press, 1991.

Weaver, Raymond. *Herman Melville: Mariner and Mystic*. New York: Doran, 1921.

Wilson, Katharina M. *An Encyclopedia of Continental Women Writers*. 2 vols. New York: Garland, 1991.

Wineapple, Brenda. *Hawthorne: A Life*. New York: Knopf, 2003.

ART INSERT CREDITS

Herman Melville. Berkshire Athenaeum
Sarah Morewood. Berkshire Athenaeum

—

Broadhall in 1900. Library of Congress
Map of Berkshire County. The National Publishing Company Railroad
   Map of New England (1900)

—

Nathaniel Hawthorne. Library of Congress
Dr. Oliver Wendell Holmes Sr. Library of Congress

—

Drawing of Arrowhead. Raymond Weaver, *Herman Melville: Mystic
   and Mariner* (1921)
View from Melville's room at Arrowhead. Author's photograph

—

Mary Butler. William Allen Butler, *A Retrospect of Forty Years, 1825–1865*
   (1911)
Dryden's poems. Author's photograph
The Whale agreement. Courtesy of HarperCollins Publishers

—

Summit of Mount Greylock. Author's photograph
Evert Duyckinck. William Allen Butler, *Evert Augustus Duyckinck:
   A Memorial Sketch* (1879)

George Duyckinck. William Allen Butler, *Evert Augustus Duyckinck: A Memorial Sketch* (1879)

—

Rev. John Todd. John Todd, *The Story of His Life* (1876)
Herman and Lizzie Melville's children. Raymond Weaver, *Herman Melville: Mariner and Mystic* (1921)
Sarah Morewood and children. Berkshire County Historical Society

—

Lizzie Melville. Raymond Weaver, *Herman Melville: Mariner and Mystic* (1921)
Herman Melville. Raymond Weaver, *Herman Melville: Mariner and Mystic* (1921)

For a detailed index of this book, and many other extras,
please visit **MELVILLEINLOVE.COM**.